EVERYDAY TECHNOLOGY

xxxxxxxxxxxxxxxxxxxxxxxxxxxxxxxxxxxxxxxxxxxxxxxxxxxxxxxxxxxxx

science.culture

A SERIES EDITED BY ADRIAN JOHNS

Other science.culture series titles available:

The Scientific Revolution, by Steven Shapin (1996)
Putting Science in Its Place, by David N. Livingstone (2003)
Human-Built World, by Thomas P. Hughes (2004)
The Intelligibility of Nature, by Peter Dear (2006)

EVERYDAY TECHNOLOGY

×××

Machines and the Making of India's Modernity

DAVID ARNOLD

The University of Chicago Press
Chicago and London

The University of Chicago Press, Chicago 60637
The University of Chicago Press, Ltd., London
© 2013 by The University of Chicago
All rights reserved. Published 2013.
Paperback edition 2015
Printed in the United States of America

24 23 22 21 20 19 18 17 16 15   2 3 4 5 6

ISBN-13: 978-0-226-92202-7 (cloth)
ISBN-13: 978-0-226-26937-5 (paper)
ISBN-13: 978-0-226-92203-4 (e-book)
10.7208/chicago/9780226922034.001.0001

Library of Congress Cataloging-in-Publication Data

Arnold, David, 1946–
Everyday technology : machines and the making of India's
modernity / David Arnold.
pages. cm. — (Science.culture)
Includes bibliographical references and index.
ISBN 978-0-226-92202-7 (cloth : alk. paper) — ISBN 978-0-226-92203-4
(e-book) 1. Technology transfer—India. 2. Technology—India—History.
3. India—Social conditions. I. Title. II. Series: Science.culture.
HC440.T4A76 2013
303.48'30954—dc23
2012050734
♾ This paper meets the requirements of
ANSI/NISO Z39.48-1992 (Permanence of Paper).

Contents

EVERYDAY TECHNOLOGY

×××

INDIA

FRONTISPIECE. A Singer trade card, dating from the early 1890s, suggesting the early use of sewing machines by women in India and the domestic intimacy of the everyday machine. Reproduced by kind permission of the Wisconsin Historical Society, Madison, WI (WHi-57879).

Introduction

In April 2011 the long-established Indian firm of Godrej and Boyce produced its last typewriter. This is likely to have been the last such machine to be manufactured in India and one of the last to be made anywhere in the world. Based in Mumbai (previously known as Bombay), Godrej and Boyce had been making typewriters since 1955, though the history of typewriters in India extends far back before that to machines imported, mainly from Britain and the United States, in the late nineteenth century. The production of the last Indian typewriter can be taken as marking the end of a technological era—the age of the typewriter—not just for India but globally. It invites reflection on the part that typewriters and other small-scale machines, many of them pioneered in the mid and late nineteenth century, have played in the making of the modern world and in the process we now think of as globalization. Although many of these global goods—bicycles and sewing machines are other examples—were initially made in the West, they came to have a profound social, economic, and cultural influence on many other parts of the world. Indeed, one could hardly speak of them as "global goods" and as being representative of "everyday technology" unless they had found a significant place in the daily lives of people not just in the West but also in Brazil and Argentina, in Egypt and South Africa, in China and India.

This book is a study of small-scale technology in India between the 1880s, when many of these new, industrially produced goods first came into use and began to find mass markets, and the 1960s, by which time they had become widely disseminated and, like the Godrej typewriters, locally produced. How and why did these machines come into general use? Who used them, who owned them, and who (eventually) made them locally? How did they affect not just economic life and productive processes but also the ways in which people worked? How did they become part of new ways of thinking—about class, race, and gender, about politics and society at large? How far did they become harbingers of technological modernity or encounter opposition as unwelcomed agents of change? To what extent does the history of a specific technology—or of clusters of interrelated technologies—become so embedded in the recipient society that the status of such commodities as global goods can assume less importance than their local uses and vernacular meanings?

This is a book about India. That is partly because India is the part of the world whose modern history I am most familiar with and from which I can most easily locate the kind of examples I am looking for. But it is also that India, today one of the world's most populous countries, was also a large and significant part of the British Empire, and empire entails discussion of issues of technological transmission and use, of socioeconomic change and political aspiration, that were distinctive, even if they were not altogether unique. Looking at India helps to decenter the history of technology and resituate it outside the familiar ambit of Western societies in which it is so often located. India is by no means the only such society for which this can be done, and in recent years studies of Japan, China, Indonesia, the Middle East, and Latin America, among others, have also undertaken a similar task. But there is a virtue in selecting one particular location for close consideration, rather than embarking on a wide-ranging comparative analysis, precisely because we can thereby better understand the global by interrogating the local. And, finally, India is now acknowledged to be one of the leading economies in our globalized world. It is all too easy today, in visiting India's megacities with their bustling traffic, high-rise office and apartment blocks, busy shopping malls, and gleaming automobile showrooms, to forget the humbler origins of India's technological transformation or the dilemmas that informed India's earlier engagement with the machine age. It is important to understand how India, now hailed as an "Asian giant," first advanced into technological modernity and in particular to see how small technologies and small machines played as significant a part as big technologies and big machines in the making of modern India.

Few would question the dominant role that technology plays in modern life across the globe. Technology, to quote Wiebe E. Bijker and John Law, "is ubiquitous. It shapes our conduct at work or at home. It affects our health, the ways in which we consume, how we interact, and the methods by which we exercise control over one another."[1] Machines are among the most evident emblems and instruments of our modernity; they, more emphatically than anything else, divide us from the technologies of the preindustrial age. Since the late nineteenth century in particular, our ideas of time and space, of body, self, and "other," have been profoundly transformed by technological innovation and by the incorporation of new and ever-changing technologies into our daily existence. And yet the functions and meanings assigned to modern technology are not everywhere the same. Identical technologies can take on vastly different meanings between one society and another, even when that technology shares a single point of origin and its physical form remains fundamentally the same. In other words, the "social life" or "cultural biography" of mechanical objects needs to be understood in context and cannot be presumed to be uniform and universal.

This is perhaps self-evident—except that the history of technological modernity has too often been presented as a single-stranded story of advancing, and indeed progressive, globalization. Modern machines are seen as having had their birth in the cultural and economic domain of the modern West. From there they were disseminated throughout the non-Western world, whether through the agency of colonial regimes or through the distributive networks of international business organizations. In such a diffusionist model the interest in technology lies with innovation and dispersal rather than with adaptation and use. All the creative processes associated with technology are presumed to lie in and with the West. Essential stages of design and development occur in the West: that is where the capital, technical expertise, and skilled labor resources are located. This leaves the rest of the world apparently sidelined and passive, existing simply as a compliant market, or, as Partha Chatterjee puts it, as the "perpetual consumers" of someone else's modernity.[2]

But, as historical scholarship has increasingly tried to argue, this does not have to be the only model of technological modernity or the only kind of instrumentality assigned to technology within wider processes of social, economic, and political change. Old technologies do not simply wither away with the coming of the new, though materially as well as ideologically they might suffer sustained attack. And, while it cannot be denied that many of the machines that most immediately capture our ideas of technological modernity—the railroad, automobile, cinema, computer—originated in the West and were first developed there to meet Western needs and Western tastes, this does not mean that their histories, once they were transferred to other societies, to other cultures and places, were merely the extension and fulfillment of their Western forms. These machines had other lives just as they had other locations. As Bijker and Law observe, "Our technologies mirror our societies."[3] Unless one takes the antiquated view that technology is an autonomous field of human endeavor and warrants study in isolation from the society around it, all technologies must in some way be grounded in the societies in which they are created, or, as is principally true in the non-West, in the societies in which they become embedded, within which they undergo adaptation, compromise, and assimilation, through which they acquire new meanings and usages. Even if the physical components remain the same, the culture of technology will, to varying degrees, differ.

But while adopting this constructionist approach—one that sees society and technology as mutually constitutive—it is also necessary to recognize issues that arise from the distinctive kind of non-Western (more especially colonial) context with which this book is concerned. Since such technological goods as bicycles, sewing machines, and typewriters were made in Britain, in the United States, or elsewhere in the capitalist, industrialized West, it was there that the primary process of their social constitution may be said to have occurred. Indians, Javanese, Egyptians, East Africans had no discernible part in fashioning the original design and basic usages of the bicycle. It arrived in their countries as the finished product of a very different, and very distant, society. Occasionally, in a vaguely orientalist echo, as in 1890s Britain, bicycles might be referred to, rather fancifully, as two-wheeled "juggernauts;" but that says more about cycling's cult status in the West at the time than any suggestion that India (or the Hindu deity Lord Jagannath) was instrumental in their creation or in the spread of the "cycle craze."[4] We should bear in mind, too, that in India and elsewhere the very foreignness of a British bicycle or an American automobile might add to the prestige of the machine and to the social kudos and cosmopolitan sheen of its owner. Indigeneity was not always a virtue; at times, it connoted second-best.

But in most extra-European societies the bicycle and similar products of modern Western technology underwent a second stage of social constitution, what Frank Dikötter has called "creative appropriation."[5] The basic form might remain the same (though even that, as in the shape of the Asian cycle-rickshaw, might in time undergo significant modification), but as cycling passed from European enthusiasts to "native" elites and indigenous masses, the status value and cultural significance of the bicycle might change. Alongside the survival and reconstitution of older technologies, there might emerge what David Edgerton has called "creole technologies" that owed their form and function to local needs, tastes, and circumstances, and did not simply replicate metropolitan norms.[6] It is suggestive of this wider process of cultural assimilation that by the early twentieth century modern machines like automobiles, motorbikes, and typewriters, along with more traditional tools of work like chisels, plows, and hammers, were venerated during the Indian festival of Ayudha Puja and daubed with sacred ash and vermillion. Even the most conspicuously alien objects might thus be incorporated into local belief and custom or be culturally reconstituted by it (though this example would be misconstrued if it were thought to suggest that ritual and religion were the *only* means by which Indians assimilated new technologies).[7] At a time when the bicycle was still a novelty in India, prizes were given at fairs and fetes for the best or most imaginatively decorated machine: where a European might dress up a bicycle as a steamship, an Indian might crowd a machine with lights and images of the Hindu goddess of plenty.[8] Around 1918, at the height of its interracial popularity, even the god Ganapati could be represented as riding on a bicycle.[9] Rather than a vehicle for the celebration of Western ingenuity, or the global replication of a single, uniform machine culture, technological modernity more closely resembled a template to which each society brought its own ingenuity and artistry, its own sense of social ownership and cultural belonging.

One could take up this discussion of the culture of modern technology in almost any part of the world outside Europe and North America—in Africa, Asia, Latin America, and Oceania. But India presents a particularly rich and challenging area for discussion. As part of the British Empire until August 1947, it was exposed to the full force of Britain's industrial might and commercial penetration, though, significantly, it was also open to trade with other European countries, with the United States and Japan. Ever among the world's most populous societies, India appeared to offer a vast market for modern machines and manufactured goods, but contemporaries were very aware of the limitations imposed by acute and widespread poverty and by what many saw as India's technological and social inertia. In part because of this apparent conservatism but also because of the connections between technology and state power, colonial and postcolonial India was a society where the role of modern technology was intensively debated. While Gandhi famously contested the desirability of most modern technology, other Indians sought to build up-to-date factories or used ingenuity and entrepreneurship to make modern technology truly Indian. Although the technology of India's craft workers and village artisans was often deemed "primitive," privileging custom over innovation, the contrast between "tradition" and "modernity" was, in this as in many other respects, greatly exaggerated. In reality, there was constant negotiation between what were presented polemically as the polarities of old and new. The debate over technology did not end with Indian independence in 1947. Indeed, under the premiership of Jawaharlal Nehru, state planning and the pursuit of economic self-sufficiency kept technological issues to the fore in public debates and government policy.[10]

But the modern machine did not enjoy an easy triumph. Twentieth-century observers employed a simple heuristic device to capture what they saw as the technological duality of Indian modernity. India, neither wholly new nor entirely old, was *both* the India of the oxcart *and* the India of the automobile. Sometimes the coexistence of old and new caused a tired resignation on the grounds that "in the land of the ox-cart one must not expect the pace of the motor-car." At other times it was invoked more indignantly to assert that in "the new age of technic" India needed to embrace modern technology and not lapse back into an arcane past symbolized by the oxcart and the spinning wheel.[11] Commonly, however, the India of the automobile was said to be perfectly compatible with the India of the oxcart since both had their appropriate place and served complementary uses. Likewise, the imagery of modern technology could infiltrate the prose of even those who did not intend to speak in favor of the machine. In his *Reminiscences* in 1917, the Bengali novelist and poet Rabindranath Tagore likened a certain Indian verse meter to riding a bicycle: "It rolls on easily, gliding as it dances to the tinkling of its anklets." It was "more like riding a bicycle than walking." In his commentary on the Bhagavad Gita in 1926, Gandhi invoked another modern machine—the typewriter—to suggest how a state of spiritual enlightenment was like a typist's instinctive knowledge of where to find the right keys without having to look each time at the keyboard.[12]

This repeated figurative use of technology in the rhetoric and reasoning of late colonial India is significant. It reflects in practical terms the sheer diversity of technologies—old and new—that coexisted in India by the 1910s and 1920s. But it demonstrates, too, the prominence given to questions of technology by Gandhi and his followers, and the cultural and existential dilemmas this created, and not exclusively among India's intellectual elite. It suggests the intensity of contemporary debates about the nature, value, and morality of modern technology, about the desirability or otherwise of its incipient hegemony. The repeated invocation of technology in speeches, memoirs, novels, even in religious tracts, signals awareness of the intrusive presence of the modern machine and its widening availability as a shared cultural commodity, an icon of everyday use and quotidian encounter. Even among those who did not own a bicycle or a typewriter, or had little personal access to a sewing machine or a gramophone, their presence—as spectacle, as something seen and heard—was undeniable. Few Indians in the 1930s owned an automobile, fewer still traveled by airplane, and yet there were not many individuals who had not seen or heard one or the other, or for whom they did not have some imaginary use. Colorful images of machines—trains, cars, planes, cameras, gramophones, sewing machines, bicycles—adorned billboards in the streets or were used in newspaper advertisements to sell soap, matches, fireworks, and cigarettes. Some even appeared among the images painted on house fronts and interiors, as in the *havelis*, the grand merchants' houses, of northern Rajasthan. By the 1930s political activists, policemen, schoolteachers, and health workers arrived in villages on bicycles and used magic-lantern slides or cinema shows to entertain, educate, or cajole their audiences. They typed reports on their visits or phoned their superiors. When Gandhi arrived to speak against modern machines he frequently did so by motorcar, his thin voice amplified by microphones and loudspeakers. Technology did not need to be big to be significant, audible, visible, and everyday. India's "new age of technic" was not just a middle-class affair, nor solely an urban phenomenon. Increasingly it was a rural phenomenon and an aspect of subaltern experience.

This book is intended as a contribution to the understanding of the culture—or, more precisely, the acculturation—of modern technology. It seeks to address the multiple understandings and experiences of technological modernity in late colonial and early postcolonial India. It aims, so far as such an act of "provincialization" is possible, to decenter the history of modern technology away from Europe, but also away from the chronicling

of British rule, toward the inner histories of India, of its intermediate groups and subaltern classes. While many accounts of
modern technology have been written—for India as elsewhere
in the once colonial world—from the standpoint of industry or
the state, this study proposes that the best (though by no means
the only) way to understand the rise of technological modernity
is by engaging with the realm of everyday perception and experience. For that purpose it uses four examples—sewing machines,
bicycles, typewriters, and rice mills—each of which represents
a different kind of technology and a different pattern of social
use, but all of which became widespread in India by the 1960s
and which, in their different yet interconnected ways, shed light
on the role of technology in the making of an extra-European
modernity.

In looking at "everyday technologies" I am consciously moving away from the "big technologies" that have, until recently,
dominated the history of technology in South Asia and many
other parts of the colonial world. Technologies such as railroads
and telegraphs, large-scale irrigation projects, and electrifica-
tion schemes were not only big in the sense of being large scale.
They required huge capital investment, directly from the state
or backed by its guarantees, involved massive environmental
appropriation and modification, and have conventionally been
thought of as exemplifying an essentially one-way "technology
transfer" between Britain or another Western power and a recipient colonial or semicolonial territory. But there were many
modern technologies—less dramatic than the railroad, more
personal than a cofferdam—that, in their seemingly mundane
insignificance, passed relatively unnoticed by the public or unregulated by the state, their presence only marginally attested
to in newspapers and photographs, or in the incidental, back-
ground material to novels and short stories. And yet, despite
their foreign provenance and lack of spectacular impact, many of
these "everyday technologies" radically transformed key areas of
Indian life, from the street and home to the jail and factory. They
frequently possessed an intimacy, a companionable association
with family life and domestic existence, which bigger machines

lacked. They weren't merely state machines and instruments of political aggrandizement.[13] Although my interest lies primarily in work-related technologies, rather than in technologies more evidently orientated toward recreation and pleasure like the gramophone, the radio, and the cinema, in any more comprehensive survey they, too, would deserve a place. My principal concern is with those technological goods which individuals outside the privileged ranks of the European and Indian elite (though these cannot be excluded) might purchase, hire, or otherwise acquire, over which they might have some sense of cultural (and not just legal) ownership, or with which they might enter into a working or antagonistic relationship.

The technical makeup of the machine is no doubt worthy of historical consideration, but my interest lies elsewhere. For me, technology serves to illuminate wider relationships, to highlight processes of social change and the exercise of political power. Through discourse and practice, through identification with race, class, community, and gender, technology (understood as object, expertise, and ideology) becomes one means among many by which a social group exercises material control and social authority over others. As an elaboration of what has been written in recent years about the subaltern classes in India and in partial response to criticisms of its shortcomings, this book is concerned to explore some of the novel forms of subalternity that emerged with, and through, technological modernity. This subaltern engagement with the machine finds expression through the artisans, laborers, and migrant workers who were displaced from their existing employment, suffered adverse changes to their working lives, or seized the new opportunities technology proffered.

Everyday technology was not a term contemporaries used. I have consciously imposed a commonality on a range of technological goods and processes that seldom came together within a single conspectus—except perhaps in the broad imaginary of "modern life." Before the 1930s even the term *technology* was rarely used. Contemporary writers more commonly invoked *science* to encompass many of the objects and activities that we would now more readily brand *technology*. Colonial authorities (if at all they cared for such things) organized agricultural and industrial exhibitions, praised power-driven "engines," spoke of "mechanical means" and "devices."[14] Gandhi's anti-industrial critique *Hind Swaraj*, written in 1909, was directed at factories and machines: it targeted "modern civilization," not "modern technology."[15] When illustrated supplements appeared in Indian newspapers and magazines in the 1920s and 1930s, showing high-speed trains, outlandish automobiles, or labor-saving domestic appliances, it was under such titles as "Science and Invention."[16] But the term *everyday technology* allows us to concentrate far more specifically on small machines and the lives of technological modernity they reveal.

CHAPTER ONE

India's Technological Imaginary

The history of technology is more than a history of material objects and physical processes. It is equally an inquiry into the exercise of the human imagination. Just as the formation of modern
nations or the conceptualization of a region as vast as India can
be fashioned by the ways in which people, individually and collectively, think about such things, so is the history of technology molded by the ways in which people identify with particular
technological goods, skills, and processes, or, conversely, seek to
distance themselves from them.[1] Technology can inform visions
of the future, shape expectations of the present, and color interpretations of the past. Technology can serve the articulation of
the self and the determination of the other.

  In a pioneering statement of technology as the imaging and
expression of imperial power, Michael Adas argues that, in the
wake of its expansion from the sixteenth century onward, Europe moved from an initially appreciative attitude toward the
technology of the non-Western world to an increasingly negative
one in the age of industry and empire. Machines became "the
measure of men," the standard by which Europe came to understand its uniqueness and superiority, and, by contrast, interpreted the backwardness and inferiority even of civilizations, like
India and China, once held in high regard. Adas offers a cogent
argument for a cultural reading of technology in the imperial
era, but his argument also suggests the possibility of alternative
readings. Indeed, one can embark on a discussion of everyday
technology in India by inverting his argument and asking not
how Europe imagined its technological other but how that other—India under colonial rule—imagined itself. Adas observes
that the "extent to which African and Asian peoples acquiesced

to European domination out of respect for the colonizers' self-proclaimed technological superiority is hard to determine,"[2] but he does not pursue the point. It is possible to shift the burden of the argument onto Indian actors without thereby ignoring the opportunities and constraints European colonialism introduced. In order to suggest the diversity and fecundity of the Indian technological imaginary, we can begin with that imaginary at its most visionary and utopian.

Technological Utopias

In 1905 Begum Rokeya Sakhawat Hossain, then a little-known writer, published a short story entitled "Sultana's Dream." In the story the author falls asleep in a chair in her bedroom and dreams that she is being taken by her friend Sister Sara on a visit to Ladyland. In this imaginary world men no longer control the state or rule the home. On acceding to power thirty years earlier, the queen of Ladyland ordered that all women should be educated and banned early marriages. Women now run the universities for which Ladyland is famous, just as they manage factories, laboratories, and observatories. With men reduced almost to irrelevance and confined in *purdah* (secluded as their wives and daughters once were), women also cultivate the fields and attend to agriculture. Wise women have brought an end to warfare through a device that concentrates the sun's rays and, directed against the enemy, renders them powerless. Having no time for idleness and quarrelling, women have turned to science and technology to control the clouds, regulate rainfall, and prevent floods: water for domestic purposes is heated by solar power and piped into every home. Women have constructed "aerial conveyances" that make roads redundant and railroads obsolete. Having eliminated traffic, the accidents and inconveniences it once caused no longer exist. The author is finally taken to visit the queen in an "air-car" propelled by two "wing-like blades" and operated by electricity . . . and then wakes up to find herself at home alone in her own room.[3]

With its systematic inversion of the customary roles of

men and women, "Sultana's Dream" has rightly been seen as a pioneering feminist tract and Ladyland as a "feminist utopia." Rokeya Hossain subsequently made a name for herself as a social reformer, especially among Bengal's Muslims, and "a boldly controversial writer on women's emancipation." But what is equally striking about "Sultana's Dream" is its insistence on science and technology as a means of achieving female emancipation and the conjuring up of a future, alternative world in which scientific knowledge and mechanical inventions prevail. Ladyland is "a utopia where science, technology and virtue work together in perfect harmony."[4] The exact source of Rokeya's inspiration is unclear, and the sultana herself confesses to Sara, as she struggles to comprehend the solar-powered gadgetry whirling around her, that her "scientific knowledge was very limited."[5] It is not difficult, however, to find it in the science-fiction novels of Jules Verne, in the utopian essays H. G. Wells published a few years before "Sultana's Dream," and in the Indian newspapers and journals of the time, with their reports of astounding new machines that could fly, talk, cook, or sing, and more generally in what has been called the "technological obsession of the Indian middle classes."[6] Remote as Ladyland might appear to be from the actuality of India (and especially of women's lives in India) in 1905, it demonstrates nonetheless that such imaginings were beginning to enter the everyday world and quotidian consciousness of the Indian people.

It would not be hard to demonstrate from a wide range of other literary and historical sources how far the machine had infiltrated, often very positively, the middle-class Indian imaginary of the period. In "Sultana's Dream" it is the visibility and utility of the technology that is most evident, but elsewhere it is the audacity, the brashness, the lyricism, and the stylishness of the machine that captures the Indian imagination. Nirad Chaudhuri, a generation younger than Rokeya Hossain, recalled hearing as a child, while staying in his mother's village in east Bengal, the sound of steamers on the River Meghna at night. "This machine-made noise," he wrote in his autobiography, "was perhaps the least dissonant sound we ever heard at Kalikutch. The siren seemed to be the voice of that gorgeous and noble river, speaking to men in the stillness of the night. There was a touch of awe in it." At a town in Mymensingh District a few years later, Chaudhuri was startled by the arrival of a European riding a noisy motorbike. This "novel object" was rapidly adopted by a young Indian as well, "so that we very soon had two of these self-advertising machines in the town."[7] The ability of the machine to advertise itself, as Chaudhuri puts it, or as one might say to advertise the self, was among the qualities that propelled it from the realms of dreams into the sphere of everyday life.

It was shortly after the first appearance of "Sultana's Dream" that Mohandas Karamchand Gandhi, still resident in South Africa but acutely aware of political unrest in India following the partition of Bengal in 1905, published his now celebrated tract *Hind Swaraj* ("Indian Home Rule"). Despite also being sometimes characterized as utopian, Gandhi's vision of a future India could not be more different from Rokeya Hossain's. Rather than machines being a means of liberation, for Gandhi railroads, trams, automobiles, and flying machines are the epitome of all that is harmful and oppressive in "modern civilization" and the soulless, avaricious modernity that threatens to overwhelm and further impoverish an India already sapped of its wealth by colonial rule. In order to save themselves from this false "civilization," Indians are urged to return to their old ways—abandon the English language, eschew doctors and lawyers, travel by bullock cart rather than by train. If the utopia of Gandhi's imagining is a land innocent of modern machines, the dystopia that threatens to engulf it is essentially machine made. As he puts it: "Machinery is the chief symbol of modern civilization: it represents a great sin." Or again: "It is necessary to realize that machinery is bad. If, instead of welcoming machinery as a boon, we would look upon it as an evil, it would ultimately go."[8]

Only one machine, the hand-operated spinning wheel, or *charka*, barely alluded to in *Hind Swaraj* but central to Gandhi's thinking from the early 1920s onward, seemed entirely acceptable. In his view it supplied the basic need for human clothing, was a source of subsistence and employment for India's poor and unemployed, and challenged the supremacy of the textile mills, half a world away in industrial Britain, that were destroying India's capacity to feed and clothe itself. Self-rule—the ability of Indians to rule themselves in both a political and a personal sense—is otherwise unobtainable. In fact, Gandhi's attitude to machines underwent some modification over the course of his lifetime, and in the mid-1920s he went so far as to accept the Singer sewing machine as "one of the few useful things ever invented."[9] He also saw some role in the late 1930s for electric power—if it could be made freely available to India's villages. But, despite these concessions, Gandhi's basic hostility to machinery remained and even intensified in the 1930s as the threat of the machine to daily life in India and internationally seemed to be increasing.

If we take these two near-contemporary tracts together— Rokeya's "Sultana's Dream" and Gandhi's *Hind Swaraj*—we are likely to be struck by the contrast between them. In style, content, and purpose they are very different works. And yet they do have in common a shared concern for technology and its role—whether as the articulation of a desired modernity or as a profound civilizational delusion—in any future India. Where Rokeya Hossain deployed a technological imaginary apparently derived from novels, news reports, and popular scientific literature, Gandhi looked instead to Western critics of industrialization, such as John Ruskin and Edward Carpenter. His vision of India, like that of Leo Tolstoy in late imperial Russia, was for a rural, decentralized society in which technology would match the human scale and measured pace of village life. For Rokeya, by contrast, machines offered a path to female emancipation and the transformation of society at large. They were an aid to such everyday tasks as heating water, growing crops, and traveling, and to more ambitious objectives, such as stopping wars or controlling the weather. But, above all, machines underpinned a radical change in gender roles. For Gandhi only the simplest, smallest, and most essential machines—like the spinning wheel, but not even the bicycle or typewriter—could justifiably belong in an India groaning under colonial exploitation and in dire need of employment, not labor-saving devices.

However, what we can discern in both Rokeya and Gandhi is the centrality of the machine and of technology generally in Indian thinking about past, present, and future. What is striking, too, is that many of the concerns expressed in this technological imaginary—about good and bad machines, to put it starkly as Gandhi does in *Hind Swaraj*—were being articulated in the decade or so before World War I, between about 1905 and 1914, a period that can be described as marking, in both imaginary and material terms, India's technological watershed. One possible explanation for this timing is the impact of the *swadeshi* movement, whose significance will be examined in chapter 4. This campaign to buy Indian goods and boycott foreign ones arose in response to the partition of Bengal by Lord Curzon's administration in 1905, though it had precedents. The *swadeshi* cause fostered a new technological awareness in which machines and the commodities they produced became a major site of contention for the Indian nationalist struggle. But Rokeya's tract predates the formal start of that movement (if only by a few months) and expresses a gendered, rather than nationalist or anti-imperialist, understanding of modern technology's transformative powers. It is more realistic to argue that, although India may not have been touched by such a compelling sense of technological modernity as emerged in Europe and North America between 1880 and 1914, it was still both elated and perturbed by the onrush of the machine age.

What is remarkable, too, in looking at "Sultana's Dream" and *Hind Swaraj* side by side, is the contrasting role assigned to the state. Where Rokeya Hossain envisages a utopia in which women have taken control of the state and the technology that sustains it, Gandhi's opposition to the modern state is almost as intense as his resistance to modern technology. This contrast touches upon one of the fundamental dilemmas of India's technological imaginary: Was the introduction and dissemination of modern technology to be largely dependent upon the operations of the state, or would it occur in defiance of the state? Did modern technology exclusively serve the colonial state, or could it be better deployed in the service of the nation?

The centrality of the state in the imagining of technology can
be exemplified by reference to a third visionary—Meghnad Saha.
Born in Bengal in 1893, Saha was trained as a physicist and had
become a seminal influence in India's scientific community by
the mid-1930s. But he also became convinced that India could
only be freed from the burden of poverty and disease by the vigorous and systematic application of science and technology. This
inevitably brought him into conflict with Gandhi. In the first
issue of the journal *Science and Culture* that he launched in 1935,
Saha wrote:

> The great success of Gandhism is due to the fact that it express-
> es genuine sympathy with the victims of an aggressive and self-
> ish industrialism, but we do not for a moment subscribe [to the
> view] that better and happier conditions of life can be created
> by discarding modern scientific technic and reverting back to
> the spinning wheel, the loin cloth and the bullock cart. On the
> contrary, we hold that if the discoveries of science are properly
> and intensively applied they will offer far better solutions to
> our bewildering economic, social and even political problems.[10]

Saha's argument about technology was also an argument
about the state. Drawing inspiration both from Soviet state
planning and from the Tennessee River Valley Authority in the
United States, Saha called for large-scale irrigation and flood
control measures across eastern India, further elaborating this
into a scheme for rapid Indian industrialization through state-
directed planning. He saw this as a practical agenda and not an
idle vision, remarking in 1943, with a further dig at Gandhi, that
India's leaders, like those of Russia before them, should "chose
the cold logic of technology over the vague utopias of Tolstoy."[11]
In 1938, at Saha's instigation, the Congress, the political arm of
the Indian nationalist movement, set up a planning committee,
under the chairmanship of Jawaharlal Nehru, to draw up a practical schema for India's economic regeneration and future self-
sufficiency. Even while India remained under colonial rule, the
prospect of using science and technology to advance national

well-being was already being systematically proposed. Saha's secular views and scientific vision of India's future underscored the appeal of modern technology, as state-led, monumental, and hugely transformative.

Improving India

The contrasting views of Gandhi and Rokeya Hossain early in the twentieth century, like those of Meghnad Saha thirty years later, were part of a vigorous and wide-ranging debate in colonial India about modern technology—technology not merely in the abstract but as an increasingly quotidian presence. To appreciate the intense, if contradictory, nature of this technological imaginary we need to backtrack to reflect on the technological ambitions and assumptions of India's colonial regime and, reverting to Adas's argument, to consider the ways in which, under colonialism, machines had become "the measure of men."

Throughout the nineteenth century the British adhered to the view that India was in urgent need of "improvement." A concept with roots in Britain's own capitalist evolution, and in its agrarian and industrial transformation, "improvement" was an empire-wide ideal, but one that had a particular moral and empirical significance for India in view of its apparent technological backwardness.[12] The poverty of India and the recurrence of widespread famine were understood as the consequence of a primitive, technologically inefficient agricultural system, one that was unable to supply the populace with its basic subsistence needs. In colonial thought as in nationalist aspiration, poverty persistently framed the problem of Indian technology, shaping the manner of its imagining as much as the material solutions proposed for its resolution. For the colonialists, this critical disposition can be traced back to the devastating Bengal famine of 1770, the creation of the Permanent Settlement in 1793, and the belief that reforming landlords (the *zamindars* created under the new agrarian dispensation) would, as in Britain, provide the impetus for a dynamic and enterprising rural capitalism. But organizations like the agricultural and horticultural societies of

India (the first of which was established in Calcutta in 1820), and the need intermittently expressed for white settlers to force the pace of technological change through a rapid injection of energy, capital, and expertise, were also part of the wider scheme of agrarian improvement.

Contemporary sources help convey the imperial spirit that lay behind the doctrine of improvement. Writing in 1835 on India's present state and future prospects, Edward Thornton informed his readers that the "prejudice, ignorance, and poverty" of Indian cultivators stood in the way of any prospect of agrarian change. India needed to look to Britain "for that impulse to improvement which from her own sons she will never receive." Indians, he averred, were a people who, partly because they had few apparent wants in terms of food, clothing, and shelter, had "slept for centuries" and were "slow in waking to active exertion." Since the country lacked a "class of men likely at present to take an interest in its improvement," it was essential for British capitalists and settlers to provide the momentum and resources.[13]

Just over twenty years later, writing in the aftermath of the Indian Mutiny and Rebellion of 1857, Harriet Martineau was equally emphatic that India's future progress was dependent upon Britain's ability and determination to bring the "practical results" of its "high civilization" into a "scene of low social condition." Only thus, she argued, would Britain and India alike benefit from a system of foreign rule in India that was quite unlike Britain's other (white settlement) colonies. The enterprise of railroad construction, begun a few years earlier, promised to bring about more than just the much-needed economic transformation of India: it would also "introduce an experience subversive of ideas and practices, which would in natural course have taken centuries to dissolve and abolish." The application of Western technology would destroy the "adored immutability," the "revered stagnation," the "beloved indolence" of the Indian people and of the Brahmin priesthood who appeared to exercise such a superstitious and reactionary influence over them.[14] Again, almost thirty years later, Sir Richard Temple lamented not just the poverty of India but its inherent lack of "improving" zeal and the practical means to effect change. "In Europe and North America," he explained, "the working power and native force of the people are multiplied by mechanical means and scientific resources. But such appliances are yet wanting for the most part to the Indians, who have been well described as being essentially un-mechanical."[15]

One could regard these pronouncements as merely evidence of imperial self-regard and self-legitimization, testament to the overweening ambitions and recurring frustrations of capitalist imperialism in South Asia. But, beyond this generalized rhetoric, lurked a quest for a more practical engagement with the materiality of Indian technology. Alongside attempts to improve Indian agriculture through the introduction and dissemination of new varieties of sugarcane and cotton, some of which were adopted by landlords and peasants, there was an ongoing preoccupation with how new or improved mechanical devices might be introduced to the Indian countryside or propagated among the artisan classes—including plows, reapers, cotton gins, spinning wheels, sugarcane presses, water-lifting devices, wheelbarrows, and scythes. Many of these projects ended in failure, but they remained part of the continuing attempt to put small but improving machines into the hands of cultivators and artisans. At the same time, the British attempted to restrict, even to outlaw, technologies such as shifting cultivation, of which they disapproved, or modes of subsistence such as those of nomadic pastoralists and itinerant hunters, which they considered primitive, unproductive, or simply conducive to crime. The British were far from neutral in their attitudes to India's existing technology, nor were they diffident about equating technological progress, as they perceived it, with their own superior civilization.

One of the principal ways in which the British sought to propagate their technological expertise to India in the second half of the nineteenth century was by means of fairs and exhibitions in what, following Bernard Cohn, one could call the "exhibition mode" of colonial knowledge.[16] Exhibitions have largely been understood as ways in which the British presented—and therefore represented—India to the wider world. Certainly, exhibitions were a near-global phenomenon in the age of industry and empire, and there is no doubting the international importance of this particular way of displaying peoples and cultures as well as commodities and machines. But it is important to recognize, too, the parallel process by which outside technology was represented to India and India was represented to itself through fairs and exhibitions in the subcontinent.

Stimulated by the success of the Great Exhibition in London in 1851, at which a large number of Indian objects were displayed, and by subsequent exhibitions in Paris, Vienna, and Philadelphia, the exhibition idea was taken up with enthusiasm by the colonial authorities in India as well. One of the first of these was held in Madras in 1855: its aim was to exhibit representative examples of Indian produce—minerals, foodstuffs, fibers, and "country medicines"—but also to stimulate "improvement" in agriculture and industry and to show that India was "no laggard" in this regard. British machines and manufactures (with a few locally made mills and *charkas*) stood alongside "raw products" that were almost entirely Indian. Hailed as "the first great exhibition for British India," the Madras exhibition was opened in February 1855 by the provincial governor, Lord Harris, and closed two months later. Although it attracted only 26,563 visitors, the attendance was said, in the circumstances, to have been "highly satisfactory and encouraging."[17] Emphasizing the state's role, Harris saw the task of "improvement in the agricultural and manufacturing industries" of India as one that would "best be promoted by a comprehensive movement on the part of Government."[18]

Indian exhibitions held between the 1850s and 1920s fell into several different categories. Some were cattle fairs, not far removed from those long held in India, but at which "improved" breeds of animals were displayed for the admiration of—and possible purchase by—landholders and wealthier peasant farmers. Others, taking agriculture as the strategic site for Indian improvement, were displays of agricultural machinery—iron plows, harrows, machines for cleaning cotton or crushing sugarcane—usually accompanied by trials to show skeptical landholders and doubting peasants how efficient and cost-effective such devices could be. In actuality, many trials proved a disaster—the machines broke down, were too expensive for locals to purchase, or were too complicated for village blacksmiths to repair. They were found to be unsuited for hard, dry soils, too heavy to be hauled by puny bullocks, or simply unable to cut, crush, grind, or excoriate the commodities intended for their use. As one Madras official sourly observed after an unimpressive display of scythes, chaff-cutters, and winnowing machines at a cattle show in 1865, "The effect of such exhibitions is much weakened if the working is not well executed."[19] Given the high cost of staging such events and the apparently negative response of the Indians who attended them (or who enjoyed the show but declined to buy the machinery), colonial officials consoled themselves with the view that India was simply not ready for rapid technological progress and should be left to its time-honored ways. State funding and official sponsorship gradually dried up.[20] And yet, as was sometimes later recognized, many of these early agricultural exhibitions were not without effect. As we will see in a later chapter, sugarcane crushing mills with iron or steel rollers, which when first exhibited in the 1860s and 1870s seemed to have little impact, were later taken up and effectively marketed by Indian entrepreneurs.

A third kind of exhibition combined "commerce" with "civilization." With local goods ranged alongside imported wares, these were closer in form and function to the great international fairs held in London and other Western capitals from 1851 onward. The grandest of these opened in Calcutta in 1883, where the exhibits included displays relating to the fine arts, education, health, science, and ethnology, as well as imported industrial goods and agricultural machinery. The exhibition attracted an estimated one million visitors between early December 1883 and mid-March 1884. In their urban grandeur, such exhibitions appeared to be far removed from the practical needs and everyday lives of the Indian masses. Indeed, it was a sign of the exhibitions' self-confident modernity that the artifacts, foodstuffs, and medicines of the people were displayed as objects of curiosity, ethnological exhibits rather than productive resources. But they simultaneously promoted the utility of sewing machines, type-writers, rice mills and other increasingly everyday mechanical objects. N. M. Patell, the principal agent for the Singer sewing machine company in India (whose innovative sales strategy we will examine in chapter 3), took very seriously his firm's partici-pation in such exhibitions and eagerly sought a prestigious gold-medal prize for the Singer machine.

The devastating famine that hit vast swathes of southern and western India in 1876–1878, causing millions of deaths, and the Famine Commission whose report followed in 1880, directed re-newed attention to the question of "improvement." Now, how-ever, there was less emphasis on local exhibitions and the dis-semination of small machines and more on the role that could be played by government departments and the deployment of their technical and scientific expertise. Since the middle of the cen-tury, especially with the coming of the railroads, there had been an increasing colonial attachment to the idea of what Manu Gos-wami has dubbed "state works." These were large-scale engineer-ing and public works projects, whose size and ambition, quite as much as their intended utility, declared their importance as expressions of the material and political power of the imperial regime and stood as "incontrovertible evidence" of its "mod-ernizing project."[21] Celebrated in contemporary photographs, in newsprint and journal articles in Britain and India, this was an imperial "technological sublime" that seemed, like its North American counterpart, to encompass the vastness and grandeur of the country and dwarf the modest achievements of precolo-nial regimes.[22] Writing in this vein in the 1890s, G. W. MacGeorge, a former consulting engineer on the Indian railroads, remarked, "At the present day no instructed person acquainted with mod-ern India would hesitate to assert that in the whole history of governments—not excluding that of ancient Rome—no alien ruling nation has ever stamped on the face of a country more enduring material monuments of its activity than England has done, and is doing, in her great Indian dependency."[23]Citing roads, railroads, telegraphs, and irrigation canals, MacGeorge claimed that "bearing in mind the vastness of the country," these were not only "truly . . . stupendous" in themselves but, when added to "the total number of individual works of exceptional magnitude and importance," the total probably surpassed what could be found in any similar expanse of territory anywhere else in the world. They could not fail, he crowed, "to leave the English name for ever indelibly printed on the soil of India."[24]

Government-led improvement and the monumentalism of state works had a lasting legacy for the way in which not only the British but also their Indian nationalist successors thought about India's technological progress. Nehruvian state planning bore something of this grandiose vision of change. There was, however, a significant colonial reaction against such massive schemes and a countervailing desire to check India's progress to a technological modernity that was deemed inappropriate, undesirable, or impractical. One of the principal exponents of this view was George Birdwood. A member of the Indian Medical Service, Birdwood systematically extolled the virtues of India's traditional crafts and looked upon the intrusion of modern machines into the Indian countryside as a kind of blasphemy, an aesthetic as well as technological disaster that risked exposing India to the worst excesses of Britain's squalid and socially divisive industrialization. In a remarkable guide to the Indian section of the Paris Universal Exhibition of 1878, Birdwood eulogized India's village crafts while expressing "great dread" at the prospect of India being swamped with Western manufactures. Writing thirty years before Gandhi's *Hind Swaraj*, but in a similar, if more aesthetic and less ascetic, spirit, he averred that machines "cannot minister to the beauty and pleasures of life." "In India everything, as yet at least, is hand wrought, and everything, down to the cheapest toys and earthen vessels, is therefore more or less a work of art."[25]

Elsewhere, Birdwood ridiculed attempts to introduce steam plows into the Indian countryside, lauding instead the humble "Mahratta plough," which he called the "perfected indigenous plough of the country, the product of three thousand years' experience." In the hands of the illiterate cultivator this produced "not once, but twice in every year," the "magical results" of abundant, unmechanized harvests. Given "the exhaustless richness of the Indian soils and the perfected science of Indian agriculture," there was no need for Western technology and scientific expertise to intrude upon the proven skills of the Indian peasantry.[26] More than this, Birdwood took the common view (in which religiosity was the mark of Indian assimilation) that Indians, having no use for, or understanding of, modern machines ended up simply deifying them. Such was the fate of the incongruous steam plow. Introduced, in a moment of technological zealotry, into an Indian village to replace the seemingly obsolescent wooden plow, it rapidly failed and fell into disuse. "As soon as it could be moved ... it was sided into the village temple hard by; and there its huge steel share was set up on end, and bedaubed red, and worshipped as a *lingam*, the phallic symbol of Siva; and there, I suppose, it stands as an object of worship to this day."[27]Birdwood's explicitly anti-industrial views represented a minority opinion among colonial officials. But the idea that modern technology was out of place in India, especially rural India, was widespread in colonialist, as in much nationalist, thought. Here was an Indian echo of what Leo Marx identified in the American context as the persistence of "the pastoral ideal" and deep-seated antipathy to the intrusive presence of the "machine in the garden."[28]

The Ethnographic Moment

In the late nineteenth century and the early years of the twentieth—at about the time Rokeya Hossain was imagining her techno-feminist utopia—there was an apparently concerted attempt by the British to represent many of India's "traditional" craft technologies in an ethnographic mode that emphasized their "primitive" nature and minimized, if not precluded, the likelihood of technological change. Through such authoritative works as Edgar Thurston's *Castes and Tribes of Southern India*, published in 1909, or still more strikingly R. V. Russell's equivalent volumes on the Central Provinces, published as late as 1916, customary skills and traditional occupations were firmly identified with particular castes or caste-like occupational groups—potters, black-smiths, stonemasons, weavers, metalworkers, carpenters, and so on. Although one can detect significant differences of emphasis between different authors and the regions with which they dealt, in the main these ethnographic works suggested that India's traditional crafts and occupations were in rapid decline or were only able to survive, paradoxically, by virtue of their very crudeness.

Occasionally, in writing of the Central Provinces, Russell gave some hint of change. Carpenters, he noted, were migrating to the cities to make "English" furniture and so earn a higher wage, just as blacksmiths were seeking nontraditional opportunities elsewhere. In many other cases, however, as with the caste of dyers, hit by the introduction of foreign chemical dyes, or that of the oil pressers, challenged by the importation of kerosene oil for lighting and cooking and by mechanized oil pressing, it was an unqualified story of decline and displacement. Typically in this declensionist techno-ethnographical narrative there were occasional moments of regret, not far removed from Birdwood's aesthetic lament for the threatened arts of village India. Thus, in speaking of the traditional tailor, Russell noted that the growing "substitution of clothes cut and sewn to fit the body for draped clothes is a matter of regret from an artistic or picturesque point of view, as the latter have usually a more graceful appearance."[29] The large number of photographs in the Russell volumes further associated these largely caste-based artisans with the distinctive tools and products of their traditional trades: they were depicted sitting or standing among the pots, oil presses, carding bows, or brass pans that typified their trades. Little distinction was made in these monochrome ethnographies between purely village artisans producing for the village community and remunerated through a share of the local produce and the more specialized, generally urban, artisans working with quality cloth and metal goods. Indeed, the assumption was that such artisans had, at least until recently, formed part of the self-sufficient Indian village community, whose cohesion was breaking down as a result of the influx of outside, machine-made goods. They did not belong in the modern, industrial age.

Thurston's approach was more varied. He expressed less interest in artisan technology than Russell and few of the photographs in his south Indian ethnography related to traditional technologies. His concerns were more with rites and rituals, with the myths of origin and the caste histories that were becoming important to the caste wars of the upwardly mobile in late nineteenth- and early twentieth-century Madras. A latent tendency toward ethnographic nostalgia was perhaps suggested by his observation about the Kadars, a tribe inhabiting the Anaimalai hills of Tamilnadu, for, being "unspoiled by education" and "happy in their innocence," they afforded "a typical example of happiness without culture."[30] At times Thurston sought to evoke in his subjects a naïve and primitive wonder at the changes that were beginning to affect even the lives of the remote hill tribes of south India as plantations of tea, coffee, and cardamoms pushed their way into once heavily forested uplands. Thus the Muduvars of the Cardamom Hills, a people who still made fire with flint and steel rather than with matches, were "becoming accustomed to quite wonderful things—the harnessing of water which generates electricity to work machinery, the mono-rail train which now runs through their country and, most wonderful of all, the telephone." An old man described to Thurston how he created wonder and envy among his community by relating his novel experiences. "I am the first of my caste to speak and hear over five miles," he was reported as saying "with evident delight."[31] But wonder and spectacle are more the mark of primitivism than of technological adaptation.

Thurston had little to say about how technological change had already impacted on the higher and middle-ranking castes of the Madras Presidency and the extent to which private lives and public occupations had been transformed by the coming of the automobile, the rice mill, and the telephone. Somerset Playne's illustrated guide to commerce and industry in south India, published only six years after Thurston's ethnography, gave a very different insight into processes of rapid and wide-ranging technological change, including among middle- and upper-caste groups, though when it came to describing "the native races" of the south his text, too, fell back on many stereotypical ethnographic images.[32] And yet Thurston did recognize the impact technological innovation had begun to make on some artisan and laboring castes—if only in consequence of European instruction. Thus the fact that a caste of Tamil smiths and masons had improved the quality of their workmanship was "entirely due to European influence." Working under Europeans as masons, carpenters, bricklayers, and smiths on railroad construction had, it seems, taught them more exacting standards of workmanship. They had come to rival English smiths in the skill and the speed of their work, and even Malas, an untouchable caste, were becoming engine drivers, turners, and valve operators.[33] One can in fact glimpse here (as in Russell's ethnography) some evidence of a lateral shift from declining old technologies to emerging new ones and the unobtrusive fashioning of a new socioeconomic status, even among low-status castes and communities.

Elsewhere, though, the picture appeared, from a technological perspective, to be even bleaker. In his description of the *Natives of Northern India*, published in 1907, the civil servant and ethnographer William Crooke gave an account of "the village and its industries" that prejudicially, if typically for the time, also doubled as an account of "criminal and vagrant tribes." This doubling up of ethnography with criminology suggested an established continuum between low-ranking village artisans and itinerant occupations, such as hunting, fowling, and well-digging, but it also demonstrated the extent to which old trades had become marginalized, their redundant practitioners driven to the margins of the law. The blacksmith, carpenter, potter, and weaver—each made his customary appearance in Crooke's narrative, only to be dismissed as technologically anachronistic. How far, Crooke asked, had these "primitive industries of the village and the vagrant" yielded to "modern industrialism"? In parts of northern and eastern India, some endured—the barber, washerman, and blacksmith were hard to replace entirely and villagers still needed the potter's cheap and disposable wares—while others, like the weaver, barely survived. "It is only the very rudest handicrafts which continue to supply the wants of the people."[34]

Overall, there is little sense in Crooke, as in other ethnographic works of the period, of adaptability, less still of innovation. It is a history of imminent loss, not incipient transformation, one that parallels contemporary Darwinian discussion of disappearing races on the brink of inevitable extinction. The underlying argument appears to be that, even if India's educated elite dreams of technological utopias, the reality is an India in which improvement has barely begun to scratch the surface of society. The effect of such ethnographies, even if not their declared intention, was to puncture the middle class's illusions of its own incipient modernity.

A Stitch in Time

There are several ways in which one could seek to overturn this negative ethnography of Indian technological inertia and obsolescence. In a series of recent studies, Tirthankar Roy has shown how many "traditional" Indian crafts—such as weaving, leather working, and brass making—were radically transformed by the opening up of new domestic and international markets and the adoption of new technological processes, and so played a "creative role" in Indian industrialization.[35] One could also cite the contemporary views of Alfred Chatterton, who as director of industries first in the Madras Presidency and then in the southern state of Mysore, argued that real technological change was already occurring in the Indian countryside in the early years of the twentieth century. It was evident to him in the adoption by peasants and artisans of a host of small, power-driven machines—such as sugar and rice mills, water pumps, fly-shuttle looms, and sewing machines. In an implied riposte to Birdwood, Chatterton wrote, "We may rest assured that there will be no opposition to the introduction of improved tools or improved methods of working if it can be clearly shown that they are real improvements. The reputation that Indians are averse to all change and are obstinately wedded to the antiquated ways of their forefathers is not justly deserved. They are conservative but they know their own business well and many of the so-called

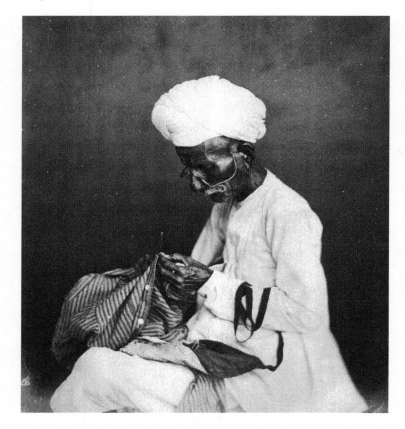

FIGURE 1.1. An Indian tailor (*darzi*) at work, in a characteristic pose, in south India, c. 1870–1880. Photographer unknown, image S0002122, reproduced by kind permission of the Royal Geographical Society, London.

improvements which they have rejected were clearly unsuitable innovations."[36] Yet another way in which we can redress the technological pessimism of the colonial ethnography is by looking more closely at one of the artisan groups commonly referred to in the "castes and tribes" literature but whose manner of work was transformed by the coming of the machine.

In the colonial iconography of Indian castes and trades, the *darzi*, or tailor, is a familiar figure. He—the iconic tailor is invariably male—sits cross-legged on a mat or the veranda of a

FIGURE 1.2. A group of south Indian tailors at work on the veranda, possibly on a European bungalow. Note the early sewing machine on the table at the rear of the picture. Photographer unknown, image S0002112, reproduced by kind permission of the Royal Geographical Society, London.

European bungalow, sewing intently or taking orders from the memsahib looming over him. He has few props—only needles and thread, scissors and thimble, and the piece of cloth or garment on which he toils alone or in a small huddle of tailors. The tailor became an exemplary image, used in ethnographic studies and exhibitions, on postcards and in photograph albums, to illustrate the country's ancient crafts and timeless traditions. Strictly speaking an occupational group rather than a hereditary caste, the *darzis* became a byword for technological inertia, the unimaginative repetition of customary skills and imitative practices. The *darzi* was deemed a copyist, not an innovator. He might be praised for being "quiet, intelligent, [and] thrifty," but more often he was branded indolent and unreliable. There was nothing "subversive" about his stitch, even if physical proximity to the bodies of female clients implied "a thousand opportunities for intrigue."[37] There was a long tradition in colonial writing of the *darzi* being bewildered by the frequent changes in European styles of dress. Asked to make a new frock by copying the design from an old garment, he does so with such unimaginative felicity that he reproduces it in its entirety, patches, darns, and all.[38]

But neither the *darzi* nor the clothes on which he worked were as unchanging as the stereotype assumed. Increasingly the *darzi* was accompanied by a sewing machine, one of the first modern machines to find its way into daily use in India as in many other societies around the globe.[39] While many *darzis*, especially in northern and northwestern India, were Muslims, there was considerable homogeneity within the community of tailors. Bombay had "Portuguese" (Goan) tailors, some in Calcutta were Chinese, and Sikhs in Punjab and elsewhere increasingly took up the trade. Perhaps because tailoring was not an occupation confined to a single caste or community, it was more accessible to newcomers than many more caste-specific trades such as weaving and metal working. Displaying spatial as well as occupational mobility, some of the migrant laborers who left India for Burma became tailors, announcing themselves on their return as "Rangoon tailors" or finding a place in the "Burma bazaar." Whether on his own initiative or, more commonly, at the behest of his European employer or the Indian master tailor who supplied him with the tools of his trade, the "traditional" *darzi* was obliged to adapt. Even his manner of work changed—often, though not always, from sitting cross-legged on the floor to working at a bench or upright at a treadle machine. The range of goods on which he worked also shifted. As new commercial commodities entered the marketplace, sewing machines were used to sew the fabric onto umbrella frames; to make curtains, bedcovers, and sails; and to stitch boots, shoes, handbags, and horse harnesses.

Indian modes of dress, too, were changing, despite orientalist representations of a timeless East. In 1867 J. Forbes Watson declared that among Indians there was "not that constant desire for change in the material and style of their costume" that was "so pronounced in Europe." "Some patterns," he continued, "which are now favourites, have been so for centuries, and cer-

tain articles of dress were ages ago very much as they now are."
Watson further claimed that many items of Indian dress, the
analogues of Western shawls and scarves, were "untouched by
needle or scissors" and were worn, unstitched, exactly as they
left the loom—as saris, dhotis, and turbans. According to him,
sewing was only introduced into India by the Muslims, from the
eleventh century onward, and hence Hindus regarded the wearing of stitched clothing as "an emblem of defeat and vassalage,
and a despotic interference with customs almost sacred from
their age."[40]

In actuality, modes of dress not only varied greatly across
India but by the late nineteenth century were also undergoing
significant change and in ways that impacted upon the utility of
the sewing machine. Male laborers began to wear shirts and cotton jackets bought ready-made in town, especially in relatively
prosperous provinces like Punjab. The calf-length skirt worn by
peasant women in Rajasthan, Punjab, and itinerant communities like the Banjaras was steadily giving way to the sari and
salwar kameez. Although experimentation with tailored jackets,
blouses, and chemises was most marked among Indian women
who came into direct contact with European teachers, doctors,
and missionaries, it became common for women, especially from
the higher castes, to seek the respectability of a stitched bodice
or *choli*. The late nineteenth and early twentieth centuries were
a time when more and more women from "respectable" communities were adapting their style of dress in order to appear in
public. As demonstrated by the "breast-cloth controversy" in the
southern state of Travancore in the 1850s, Christian missionaries
were determined to defy caste taboos by insisting on the right
of female converts to wear blouses and other forms of upper-
body clothing. Half a century later, Hindu and Sikh reform movements zealously targeted the immodesty of women who failed
to appear in public in suitably demure body-covering attire.[41]
Unlike in Japan and some other parts of Asia, the Indian tailor
and his sewing machine did not have to await the acceptance of
Western dress to find a new role and a new market.

Fashion-conscious and house-proud European residents in

India employed their own tailors, dressmakers, and milliners, some of whom were poorer Europeans or Eurasians; but innovation did not end there. The use of petticoats, shirts, uniforms, and tunics proliferated among Indians employed by Europeans too. Apart from the elaborate livery in which many Europeans dressed their household servants, the army (which had its own regimental tailors), the police, business houses, municipalities, railroad and tram companies, the telegraph service, and the post office all required their employees to dress in uniform and so needed tailors to make and mend them. Taken together, this private and public demand created unprecedented work for tailors and their machines. Part of the impetus for technological change at the everyday level thus came from within a colonial society that was undergoing far-reaching processes of socioeconomic change. Rather than look globally at how new technologies were disseminated it can be more informative to understand who required such novel goods and how new technological usages were introduced to meet local needs and aspirations.

×××××××××××××××

By the interwar years India's technological imaginary encompassed a great diversity of different ideas and perspectives. Whether among colonialists, nationalists, or feminists, there was no single view of India's technological requirements or of the solutions modern technology might offer to a society widely condemned as backward and impoverished. And yet, for all this diversity, there was by the late nineteenth and early twentieth centuries a heightened sense of both the potentialities and the limitations of technological change. Despite the continuing appeal of large-scale "state works" and an enduring belief in state-directed "improvement," the dominant colonial view was one of pessimism—that India, unaided, was too poor, inert, and conservative to be capable of the kind of technological change that had transformed, and was continuing to reinvigorate, Western societies. If some observers like Chatterton ardently believed that small-scale machines were reshaping the countryside, others like Birdwood were skeptical of the benefits Western technology could ever bring to rural India, or, like the colonial ethnographers, they emphasized the demise of "traditional" artisans rather than seeing in them any significant capacity for innovation. Indians might display more technological optimism than their colonial rulers, whether in the form of Rokeya Hossain's feminist utopia or the state planning and nation-serving technology envisaged by Meghnad Saha. Some Indians were evidently excited by the prospect of mechanical change, while others like Gandhi took a more guarded, even hostile, view. All these diverse perspectives underscore the centrality of technology to the wider imagining of India and its modernity, but, taken as a whole, they show a negligent tendency to ignore or downplay the importance of those small-scale and increasingly everyday technologies that were beginning, like the tailors with their sewing machines, to change the lives and the livelihoods of so many Indians.

CHAPTER TWO

Modernizing Goods

From Transferred Goods to Acculturated Technologies

Beginning in the late nineteenth century and increasingly by the 1920s and 1930s, India, like many other parts of the non-Western world, became a market for an entirely new set of industrially produced consumer goods and small-scale machines. Among the most conspicuous and widely disseminated of these "modernizing goods," ranging from the domestic to the light industrial, from the utilitarian to the recreational, were sewing machines, gramophones, typewriters, bicycles, cameras, clocks, and watches.[1] Before 1914 hardly any of these machines were made in India, though they were in many cases assembled and repaired there. They can be understood as examples of the way in which industrialization in Britain, western and central Europe, and the United States was creating global markets, and in the process forging new consumer tastes and founding the regional reputation of international brand names like Singer, Raleigh, Kodak, and Remington.

It is possible, too, to see the history of these mechanical commodities (along with earlier and grander ones like railroads, steamships, and telegraph systems) as examples of the "transfer" or "diffusion" of modern technologies outward from the West to the less economically advanced and technologically privileged regions of the world. In most cases, though, it was the products of new industrial processes that were disseminated, or the basic skills required for their operation and use, rather than the actual ability to make the goods or develop the industrial processes required for their local production. Following this line of argument, the social construction of technology accordingly occurred elsewhere, and not in countries like India, which were the recipients of machines whose design and utility had been established in and for very different societies. With respect to India, as in relation to many other colonial or semicolonial societies, the process of "technology transfer" has been directly linked to imperialism. Imperialism created the physical infrastructure and socioeconomic conditions through which new mechanical commodities could be introduced and find markets. Or, more critically still, imperial control and the favored status of foreign firms precluded or inhibited local production. Unlike Britain or the United States, colonial India lacked a substantial small-arms industry to provide the entrepreneurial and technological foundation for the manufacture of sewing machines, bicycles, and, latterly, automobiles. Nor, under colonialism, did India have the political freedom to create its own tariff walls and so exclude foreign manufactures. Like locomotives, steamships, and firearms before them, such novel commodities as sewing machines, bicycles, and gramophones long remained imported goods. The profits from their sale, and, no less significantly, the technological expertise gained through their manufacture, accrued elsewhere.

A technology transfer argument helps to explain how commodities like the sewing machine and bicycle came to India when and how they did. It might also suggest some of the reasons why, under imperialism, indigenous production struggled to compete with imported goods. But, except in its subtler forms, the transfer argument provides little explanation as to why some imported machines became widely adopted while others did not. It fails to explain why some social groups took to their use, and others didn't, or how certain kinds of "modernizing goods" acquired a cultural, social, and political significance and so, despite their foreign origins, became integral to indigenous ways of thinking, working, and being. And, given that in India it was often, by global standards, the paucity rather than the plentitude of modern consumer goods that was so striking, a technology transfer argument does not in itself account for the relatively rapid process of cultural absorption as well as technological assimulation. What is needed, therefore, is a "socio-technical" history that sees society and technology as part of the same "seamless web," mutually connected and continuously interlinked, but which is sufficiently flexible to accommodate the derivative nature of the technologies involved.[2] Technology cannot meaningfully be understood, I would argue, without considering the society in which it becomes embedded, even when the technological goods themselves remain largely foreign. A process of local acculturation needs to be recognized, in which the uses and meanings of machines can be seen as both conforming to, and being transformative of, the recipient society.

The Coming of the Sewing Machine

In recent years the history of the sewing machine has attracted wide historical interest—not only for North America and western Europe but also for Latin America, the Middle East, and South and East Asia. This is unsurprising. Light and easy to use, the sewing machine was one of the first machines to find a global market. Its cheapness and mobility enabled its use in factories and households but also in relatively remote rural areas. One of the first machines to enter the home and make use of women's labor, the sewing machine had a near-global impact on work practices and fueled the rise of the modern, and increasingly global, garment industry. Although sewing machines were made in Britain, Germany, and other countries as well, until the 1930s the international trade was dominated by American manufacturers, especially Singer which also introduced new forms of business organization and pioneered novel sales techniques. A commodity made possible by modern industrial mass production, the sewing machine was also the beneficiary of the post-1870s expansion of world trade, the network of steamships, railroads, telegraphs, postal services, and newspapers that facilitated its rapid dissemination and growing popularity. Emblematically modern, the sewing machine in turn encouraged the adoption of other small-scale domestic and light industrial machines. Like the bicycle, it became one of those items of consumption emerging economies—like Japan, Brazil, India, and more recently China—sought to replicate and manufacture for their own or international markets. Given its international history, the sewing machine offers a primary example of a history that is both technological and social, and one which, as well as illustrating global patterns of dissemination, demonstrates, no less emphatically, how societies responded differently to the same technological good.

Sewing machines were first introduced into India in the late 1850s (the same decade that saw the birth of its railroads), but they only began to arrive in the country in significant numbers in the 1890s. Although, in discussing global sales of Singer sewing machines, Andrew Godley remarks that sales in India "remained insignificant" in the period up to the end of World War I and by 1916 had barely reached even 1 percent of Indian households, the impact of the sewing machine, socially as well as commercially, was far greater than this meager statistic would suggest.[3] Despite the economic and social constraints, including "arrested development" and endemic poverty, which accounted for low levels of Indian demand, between 1900 and 1950 more than two million sewing machines were imported into India (see tables 2.1 and 2.2). While the Indian market was not as large and receptive as those of many European or even Asian countries like Japan and the Philippines, imports quadrupled between 1900 and 1914, underscoring the importance of this period to technological change in India. Despite the partial boycott of foreign goods caused by the *swadeshi* movement and the impact of World War I on global manufacturing and trade, the volume of imported machines continued to rise until 1916–17, before declining sharply with the postwar trade slump in 1921. As with other imports, sewing machines reached a high point in the late 1920s until the Depression, which halved India's foreign trade between 1928 and 1933, brought a further fall. Imports rose again in the mid-1930s reaching a peak in 1937–38, when just over ninety thousand machines entered India. The contrast with Japan is instructive: there Singer sales also peaked in 1937 but, faced with indigenous competition, began to fall rapidly thereafter.[4] In India there was as yet no local competition. Imports dropped during World War II but reached new heights in the late 1940s and early 1950s. Thereafter, stiff import duties and the indigenization of sewing-machine production finally began to impact on the sale of foreign machines in India.

Although slow to enter the Indian market—Singer's first machines were not sold there until 1875, twenty years after its operations began in the United States—the Singer Sewing Machine Company, America's "first international company," made the vast majority of sewing machines sold in India during the colonial period.[5] Sewing machines formed part of that array of goods—from clocks and cash registers to typewriters, electric fans, and telephones—entering South Asian markets by 1914 that came not from Britain, the colonial power, but from the United States and that drew primarily upon American rather than British or European technical expertise. Under British India's free-trade regime, it was possible for American goods, like those of other industrial nations, to compete with and substantially outsell those made in Britain. In this respect, conventional ideas of a British Empire dominated by British goods need to be revisited and the extent of America's cultural, commercial, and technological penetration of India by the 1930s more fully recognized. However, from the early 1880s most of the "American" sewing machines reaching South Asia came from Singer's Clydebank factory near Glasgow and so entered Indian trade statistics as British manufactures. As early as the 1850s Singer had recognized the limits of US domestic demand and systematically set out to capture overseas markets. Initially the company faced tough competition from such British firms as Willcox and Gibbs, whose machines appear in many photographs of European domestic scenes in late nineteenth-century India. It also competed with the American company Wheeler and Wilson, which was the first sewing-machine manufacturer to establish itself in India, but which was eventually taken over by Singer in 1907.

By the early 1900s Singer machines dominated the South Asian market, as they did most other parts of the world. However, Singer never entirely excluded competitors like the Ger-

FIGURE 2.1. The sewing machine was one of the few modern machines to which Gandhi gave his approval. Here a Gandhian volunteer tailor is at work on a pedal-powered German Pfaff machine in 1946. Photo by Margaret Bourke-White, reproduced by kind permission of Getty Images (50878431).

man maker Pfaff, which began producing sewing machines in the 1860s, and which by the 1920s and 1930s was energetically seeking markets overseas, including Brazil and India. In 1900, 74 percent of the sewing machines imported into India were British-made compared to 5 percent each from Germany and Belgium, and less than 1 percent from the United States. On the eve of World War I the British (mostly Singer) market share had dropped to just over 66 percent while the German portion had risen to almost 32 percent. In the 1920s and 1930s the British and German shares remained at roughly the same level, about two-thirds and one-third, respectively. World War II again eliminated German competition and in the late 1940s, when almost 90 percent of the sewing machines imported were from Britain, Singer held "a semi-monopolistic position" in the Indian market.[6] With its foreign rivals removed, Britain in 1948–49 exported nearly fifty-one thousand sewing machines to India, while the United States, Germany, Sweden, Japan, and Italy together sent barely four thousand. Although dwarfed by machine imports for the textile and railroad industries, this was, and long remained, a lucrative trade: in 1913–14 the value of imported sewing machines was put at £238,805; by 1920–21 it had reached £552,565 (with a further £128,013 in sewing-machine parts). In 1937–38 imports were worth Rs 6,665,809.*

Thereafter, the position began to change. The ready-made garment industry developed rapidly in India in the 1960s and 1970s: German and Japanese competition, especially in the market for industrial sewing machines, grew at the expense of British and American makes. In addition, Indian manufacturers, helped by new import controls, began to win a dominant share of the domestic market and to export locally made machines to other Asian and African countries. As the number of sewing machines in operation rose between 1950 and 1961 from 145 to 742 for every million of the population, they, like bicycles and radios, became one of the key indices used to indicate improving living standards in India.[7]

Until the early 1950s, Indian manufacturers presented little threat to foreign manufacturers. Three small-scale sewing-machine producers were established in the late 1930s: Jay Engineering, which made the "Usha" brand in Calcutta; the Delhi Sewing Machine Company, whose first machines were marketed in 1938; and the Indian Sewing Machine Manufacturing Company in Lahore. But all three remained heavily dependent on imported components (including such essential items as steel needles) and they struggled to compete with such well-established makes as Singer and Pfaff in terms of quality and after-sales servicing. Although during the war India had been "starved" of sewing machines, with supply falling far short of demand, restrictions on materials, like high-grade steel, inhibited indigenous production, and factory machinery was diverted to making munitions

*In the nineteenth century, an Indian rupee (Re) was officially valued at a tenth of a pound sterling (£) or 2 shillings. By the mid-1920s its value had fallen to 1s 6d and by the mid-1930s 1s 4d.

TABLE 2.1. Annual imports of sewing machines, bicycles, and typewriters into India, 1900–1901 to 1951–52

Year	Sewing Machines	Bicycles	Typewriters
1900–1901	15,251	n/a	n/a
1901–2	15,415	n/a	1,233
1902–3	18,770	n/a	1,643
1903–4	26,247	n/a	2,363
1904–5	26,455	n/a	2,600
1905–6	26,683	n/a	2,507
1906–7	29,207	n/a	3,794
1907–8	28,241	n/a	3,421
1908–9	27,642	n/a	4,216
1909–10	29,291	n/a	3,395
1910–11	41,065	n/a	3,458
1911–12	39,797	n/a	4,152
1912–13	57,581	29,150	5,677
1913–14	61,183	34,577	6,267
1914–15	51,149	23,904	5,237
1915–16	40,357	19,342	4,493
1916–17	74,642	15,309	5,521
1917–18	57,761	10,937	8,380
1918–19	27,534	7,825	6,947
1919–20	48,882	21,912	11,559
1920–21	62,964	46,706	16,888
1921–22	23,845	6,315	7,364
1922–23	46,050	22,940	5,103
1923–24	48,385	44,884	7,930
1924–25	61,229	62,003	7,445
1925–26	70,835	96,649	10,947
1926–27	71,497	115,528	13,790
1927–28	75,264	138,783	16,952
1928–29	78,241	163,432	21,487
1929–30	68,680	142,052	15,774

TABLE 2.1. (continued)

1930–31	52,222	53,472	15,544
1931–32	45,317	49,672	7,275
1932–33	39,109	69,528	3,944
1933–34	54,886	88,624	5,827
1934–35	83,354	106,286	14,925
1935–36	84,755	133,595	15,404
1936–37	64,791	165,390	15,124
1937–38	90,023	170,664	19,721
1938–39	61,231	138,036	11,877
1939–40	75,382	92,249	14,480
1940–41	34,586	50,222	15,450
1941–42	41,011	54,455	14,367
1942–43	18,575	16,134	7,685
1943–44	4,389	25,356	2,340
1944–45	10,050	37,391	4,773
1945–46	28,359	76,141	30,841
1946–47	59,486	212,554	n/a
1947–48	53,385	261,958	24,561
1948–49	56,938	264,392	42,730
1949–50	49,043	268,373	25,397
1950–51	23,425	165,811	18,536
1951–52	28,872	283,127	20,575

Source: Annual Statement of the Sea-Borne Trade and Navigation of British India for the relevant years. Data after 1936–37 exclude Burma, and after 1946–47 Pakistan.

and war materiel. When production resumed in 1946–47, fewer than eight thousand sewing machines were made in India, rising to seventeen thousand by 1948–49, still less than a third the number of imported machines. It was only in the late 1940s that Indian manufacturers were in a position to expand and profit from the active support of the new government of India.

But a history of technology that isn't also a history of the

TABLE 2.2. Imports of sewing machines, bicycles, and typewriters into India by decade, 1900–1949

Year	Sewing Machines	Bicycles	Typewriters
1900–1909	24,320	n/a	2,517*
1910–1919	49,995	16,296**	6,169
1920–1929	60,699	82,929	12,368
1930–1939	57,965	106,751	12,412
1940–1949	35,378	126,698	16,814***

Source: Annual Statement of the Sea-Borne Trade and Navigation of British India for the relevant years. Data after 1936–37 exclude Burma, after 1946–47 Pakistan. * Excludes 1900–1901. ** Excludes 1910–1911. *** Excludes 1946–1947.

people who use it doesn't make much social sense. Despite the kind of colonial rhetoric noted in the previous chapter, in which Indians were seen as averse to innovation and suspicious of new technology, there is little evidence of significant cultural resistance to sewing machines. Indeed, despite their small numbers before 1914, they passed almost effortlessly into everyday use. In India, as elsewhere, such products of modern science and technology "swiftly moved," as James Masselos has put it, "from [being] objects of wonder to objects of use; they became part of the daily round of living. Neither social constraints nor the force of custom inhibited significant use of the achievements of the nineteenth-century world."[8]

This process of assimilation will be further elaborated in the following chapters, but it can be noted here that the sewing machine was a remarkably successful technological implant, a readily accessible machine that required very little capital investment or prior acquaintance with machinery to speed its introduction. Many "traditional" tailors were illiterate and relied on their employers to provide them with their machines. A detailed study of darzi family budgets in Madras city (now Chennai) in 1938 showed that it required only Rs 50 to set up a tailor's shop. It cost about Rs 200 (the equivalent of ten months' income for a tailor) to buy a new machine, but one could be purchased second-hand for far less or hired for between Rs 20 and Rs 50 a month.[9] Alfred Chatterton, whose enthusiasm for small machines has already been noted, remarked shortly before World War I that Singer sewing machines were to be found "in almost every tailor's shop in the country." Although they were "somewhat delicate and complicated" pieces of machinery, he observed, "the facilities for the repair or renewal of parts have been so widely diffused that the tailors find no difficulty in keeping them in working order."[10]

Without the fanfare that greeted the coming of the railroad, the sewing machine was widely, if sometimes cautiously, welcomed as a harbinger of change, reaching into even the remotest Indian villages.[11] However, as with much else about the dissemination and reception of everyday technology, the evidence is fragmentary and often anecdotal, but works of fiction provide one useful indicator. Thus, in Rudyard Kipling's novel, when Kim first encounters a gramophone he is able to recognize it as being "some sort of machine, like a sewing machine." It looks and smells rather like the sewing machines he already knows from the backstreets of Lahore.[12] The visibility and audibility of the sewing machine in bazaars and roadside tailors' shops made its presence, and the dexterity of the tailor who worked it, a familiar trope in fictional accounts of Indian life. In Mulk Raj Anand's 1930s novel, *Untouchable*, Bakha, the uneducated untouchable of the title, is said to be absorbed in wonder at the "manipulation of a sewing-machine" by a tailor in the bazaar.[13]

During their tour of India in 1912, Sidney and Beatrice Webb encountered sewing machines in the classroom of a Hindu girls' school in Allahabad and in Peshawar in the North-West Frontier Province, where tailors sat on the floor of their tiny shops, "working Singer's sewing machines."[14] In cities such as Calcutta (Kolkata) and Dacca (Dhaka) large tailoring establishments sprang up to meet the demand for ready-made shirts, coats, and uniforms. Sewing machines found a particularly lucrative market in Punjab, especially in Lahore and Amritsar. Across the province as a whole the number of those employed in tailoring and allied occupations rose by 40 percent between 1901 and 1911. In Lahore, the largest city in Punjab, with its rail workshops and army cantonment, the local demand for uniforms spurred the growth of clothing factories.[15] Such enterprises anticipated the later rise of the South Asian garment-making sweatshop. In this respect, the impact of the sewing machine on the mass production of cheap clothing was not very different from the Western experience, though it occurred rather later.[16]

Nor was it only town tailors who took to sewing machines. In a still mainly rural society, penetration into the countryside remained an important marker of the effective dissemination of any modern machine. It was, therefore, significant that in Punjab in 1911 there was "hardly a tailor now without a sewing machine. Even in the villages a tailor would beg, borrow, or steal to equip himself with a cheap machine, and, if he cannot find enough customers in one village, he will rather set apart a certain amount of time for regular rounds and attach himself to a group of villages."[17] *Darzis* and their machines soon became a common sight—and sound—across rural India. In 1927 Margaret Read noted, "Almost every village tailor now has a sewing machine, whose tick-ticking makes a strangely alien sound in the street where the thick dust muffles all sound of footsteps, human or animal."[18] She was not alone in find odd the presence of a "town Darji" at work in the village, "amid surroundings of apparently the most incongruous simplicity."[19]

The Advent of the Bicycle

Like sewing machines, bicycles were a pioneering technology. As elsewhere in the world, they met the growing demand for a cheaper, simpler, and more autonomous mode of individual transport than horses, carriages, and railroads could provide. They gave to men, women, and children a new sense of mobility, created the possibility of increased sociability, and, until their popularity waned, offered a new public stylishness. The durability and versatility of the "safety bicycle" by the close of the nineteenth century allowed it to be used for work as much as pleasure, for country roads as well as urban streets. The recreational and therapeutic value of cycling was widely touted. In what became a near-universal craze, cycle clubs and cycling magazines sprang up almost everywhere. The bicycle in turn provided the pioneering technology and demonstrated the market potential for the subsequent rise of the automobile and motorbus.

India was not immune to these developments. Even though the manufacture of bicycles required a more advanced engineering industry than existed in South Asia before the 1930s, India imported large numbers of bicycles and had the capacity to assemble and repair machines. There, as elsewhere, bicycles were the beneficiaries of the urban and rural road network that had been developed under colonial rule and was greatly extended after 1900 with the coming of the automobile and motorbus. By the 1920s Indians riding bicycles were common sights in cities, towns, and villages, traveling to and from work, carrying friends and family members on the crossbar or rear carrier, hawking goods, or, as one still sees today in India as in much of South and Southeast Asia, transporting all manner of bulky commodities—sacks of grain, kerosene cans, old tires, timber, bricks, and baskets of chickens.

Bicycles began to enter India in significant numbers in the 1890s but were only entered separately in government trade returns in 1905 and then at first by value, not quantity. In 1905–6 bicycle imports were worth Rs 1,615,720, with the bulk of machines arriving from Britain. The figures for the forty years from 1912–13 to 1951–52, given in tables 2.1 and 2.2, track much the same erratic progress for bicycle imports as for sewing machines. In the years following World War I (as sales of automobiles and motorcycles also increased) bicycle imports rose substantially. In 1920–21 roughly three times as many bicycles as automobiles were imported into India (47,000 to 15,000), though bicycles were worth only a fraction of the value of automobiles. In numerical terms, in the mid and late 1920s bicycles began to overtake and then outstrip imports of sewing machines, peaking at 142,052 in 1929–30. Thereafter, like other imports, they slumped in the early 1930s but recovered strongly to 170,664 in 1937–38. Before

1914 new bicycles were relatively expensive, costing between Rs 85 and Rs 250, but during the Depression the falling price of bicycles, in many cases dropping below Rs 50, made them increasingly affordable. World War II brought cycle imports almost to a standstill, but they surged again after the war, reaching unprecedented heights in the late 1940s and early 1950s. By that time the cost of purchasing or maintaining a bicycle had become a standard part of lower middle-class household budgets.[20] Overall, between 1912 and 1946 around 2.5 million bicycles were imported into India, on average about 70,000 a year, but a further 1.2 million bicycles entered India in the years between 1947–48 and 1951–52, equivalent to more than 200,000 a year.

Of the four technologies taken up in this book, bicycles have perhaps the strongest claim to being items of mass consumption. It is indicative of increasing Indian use that Dr. Aziz in E. M. Forster's 1920s novel *A Passage to India* first appears riding a bicycle. An article in the *Illustrated Weekly of India* claimed in April 1939 that there was now "hardly a village in India where at least one bicycle is not in use. As a cheap means of transport the bicycle has come to stay."[21] And yet, given the poverty of India and the size of its population, the same caveats apply to bicycles as to sewing machines. Even in the late 1940s the ratio of bicycles to population in India was much lower than in most Western (and many Asian) countries. It was calculated in 1946 that of the world's 70 million bicycles, India, despite having around 400 million people, had only 1.5 million bicycles. This was equivalent to less than four bicycles for every 1,000 Indians, compared to 255 for every 1,000 inhabitants in Britain, 463 in the Netherlands, and 539 in Denmark. However, if we take into account the concentration of bicycles in urban areas, and the predominance of adult male users, perhaps as many as one in every ten men in a city like Madras in the 1930s owned or used a bicycle.[22] The 1950s saw a further leap in bicycle use across India, with cycles available for hire even in small towns and villages. Perhaps it is only at this point that one can accurately speak of India having entered the "cycle age." However, in 1960 when Jawaharlal Nehru, the prime minister, used that phrase, adding that the bicycle had "invaded the villages" and become "a very popular means of transport all over India," he did so with evident regret that India was still a long way from joining advanced industrial nations in the age of electronics, jet travel, and atomic energy.[23]

Unlike some everyday technologies (such as typewriters, which were largely imported from the United States), until the 1960s the great majority of India's bicycles were made in Britain. They came from a number of different manufacturers, such as Hercules, BSA, and Rudge-Whitworth, but by the 1940s Raleigh had emerged, like Singer among sewing-machine manufacturers, preeminent, partly through a similar process of absorbing smaller rivals. There was competition in the interwar years from European makes and cheap imports from Japan. A fully-equipped Raleigh bicycle in the mid-1930s cost Rs 60, and other British or locally assembled machines between Rs 40 and Rs 50, but a Japanese bicycle could be bought for as little as Rs 19. And yet, despite their cheapness, Japanese bicycles never acquired the prestige of British makes and were considered less durable in Indian conditions: they peaked in the mid-1930s at barely 8 percent of the import market.

Not all technological goods, however, were acquired or gained circulation through purchase or by means of loans. Bicycles were supplied by employers for office workers and low-ranking government servants, such as policemen, postmen, telegraph boys, and sanitary workers. Poorly supplied with public transport, the vast expanses of the new imperial capital, New Delhi, necessitated the grant of bicycle advances and loans for government employees to get to and from their places of work. Bicycles were also stolen or obtained by fraud in significant numbers. In Madras city in the mid-1930s cycle thefts averaged around 230 a year; by 1938 that number had risen to 440 (admittedly only a tiny percentage of the estimated 33,000 bicycles in Madras at the time). Bombay and the cities of northern India were similarly plagued by bicycle thieves, with Bombay alone reporting more than a thousand thefts in 1939.[24] In some places organized gangs of bicycle thieves were at work; in others cycles taken out on loan were never returned. Despite the efforts of the police, few thieves

FIGURE 2.2. Since their introduction more than a century ago, bicycles have proved their value transporting goods as well as people. Here a laborer in present-day Thanjavur in south India carries sugarcane to market. Author's photo.

and fraudsters were detected, and most stolen bicycles quietly disappeared into the anonymity of the backstreets, suburbs, and villages.

The bicycle and its components lent themselves to a variety of additional uses. Bicycles were adapted by local blacksmiths to make three- or four-wheel carts for transporting and selling goods as well as to make cycle rickshaws. In the 1920s one enterprising Calcutta firm offered for sale a "patent water cycle" whose only use appeared to be for duck shooting.[25] The bicycle's saddle, pedal, crank, and chain could provide the mechanism for an experimental foot-powered loom,[26] and one can still see today, on the streets of Mumbai or Ahmedabad, a knife grinder who raises his bicycle onto its stand and then uses the pedals and crank to turn a grinding wheel attached to the frame. The bicycle became a font of further local inventiveness.

The Immigrant Typewriter

Of the four main technologies examined in this book, typewriters might appear least eligible for consideration as "everyday" objects. They might be regarded, with some justification, as a technological good almost entirely confined to urban offices and literate elites, remote from popular technological encounters. Certainly typewriters were not cheap: in the late 1930s a new Remington machine might cost as much as Rs 316, several times the asking price of a bicycle.[27] However, there are a number of reasons for the inclusion of typewriters. Along with sewing machines, bicycles, cinema, cameras, and gramophones, typewriters formed part of a series of innovative technologies developed in the second half of the nineteenth century that served, on a global scale, to articulate a new sense of modernity. As an advertisement for the Chicago-made Oliver typewriter insisted in 1904, "All progressive people should see this machine."[28] Although India did not (for reasons we will see shortly) fully share the sentiment, the typewriter was widely seen internationally as an exemplification of modernity for its speed and convenience, for its mobility and its bureaucratic utility, and as a vehicle for artistic self-expression.[29] Just as the bicycle introduced the foot pedal as a means of propulsion, so typewriters pioneered the manual keyboard and so helped pave the way for the computer. Typewriters are still used in India to test the keyboard skills of some applicants for government jobs.

The typewriter did not serve the creative needs of individual writers alone. It was also at the heart of the modern office, that "oasis of modernity" as Dikötter describes it in relation to modern China.[30] There it functioned alongside a cacophony of other machines—telephones, electric fans, duplicators, calculators—that transformed bureaucratic work regimes while demanding new operational skills. In many parts of the world, the typewriter brought women into the modern workplace or helped them acquire a new status as writers, journalists, organizers, and activists. Sadly, the typewriter's social and technological impact is less well documented for India than for most Western countries. Histories of Indian business houses pass over the coming of the typewriter with such unhelpful observations as: "The revolutionary effects of this apparatus on office routine need no description or comment." And yet it is clear that in cities like Calcutta, Bombay, and Madras the typewriter, like the automobile and the electric tram, signaled the "advent of a new era."[31] Just as home, street, and factory became prime sites for many other new technologies, so the typewriter made its noisy appearance in government and business offices, homes, clubs, welfare societies, and political organizations, transforming the ways in which novelists, journalists, politicians, and administrators pursued their daily work.

We are accustomed in the West to think of the typewriter as an indoor object, but it is still possible to see in India groups of typists sitting outside post offices and law courts, perched on stools in front of now antiquated machines, bashing out petitions, letters, and affidavits. In his novel *A Fine Balance*, set in Bombay in 1975, Rohinton Mistry describes typists outside the city's high court, "sitting cross-legged in their stalls before majestic Underwoods as though at a shrine, banging out documents for the waiting plaintiffs and petitioners."[32] Forty years earlier, in 1936, the *Illustrated Weekly* complained about typists who had "planted themselves on the pavement" outside Bombay's central post office, obstructing passersby and forcing them to walk in the gutter. It was even alleged that the typists slept on the sidewalk alongside their machines and had their mail delivered there.[33]

Observing pavement typists today, one gets a sense of the way in which the typewriter and other machines that became denizens of the Indian street or functioned along its margins served as the site of a new sociability, as the typist, the cycle repairman, the tailor, or the rice-mill operator chats with his (less likely her) customers or fills idle time gossiping, smoking, and drinking tea.

The volume of typewriters imported into India never matched that of sewing machines and bicycles. Between 1910 and 1950 about half a million typewriters reached India, a quarter of the number of sewing machines. It is impossible, though, to quantify the additional number of typewriters, especially portables, which entered India as personal effects. Before World War I the number of imported machines was small—barely a thousand in 1901–2 and rising to only 6,267 in 1913–14. As with other mechanical imports, the war restricted supply. Once it was over, there was a surge in imports in 1920–21, before demand fell back and only gradually recovered over the course of the decade. In 1928–29 typewriter imports stood at 21,487, the highest figure reached until the end of World War II. Even so, despite a dip in the early 1930s as the Depression hit India and forced retrenchment in government offices and commercial firms, typewriter imports remained buoyant for most of the 1930s until war again intervened.

More remarkable than the scale of typewriter imports was their provenance. In 1906–7 the bulk of machines came from Britain: of the 3,794 imported in that year, 59 percent came from Britain, 39 percent from the United States, with barely a dozen from the next nearest supplier, Germany. But already by 1909–10 American machines had begun to outstrip British makes: 59 percent of the 3,395 typewriters imported that year came from the United States and only 39 percent from Britain. As with automobiles, World War I, which saw British and European manufacturing diverted to military needs and caused serious disruption to international shipping, strengthened the hold of the United States on the Indian typewriter market until it became a virtual monopoly. In 1915–16 three-quarters of India's typewriters were imported from the United States, and in 1920–21, as post-

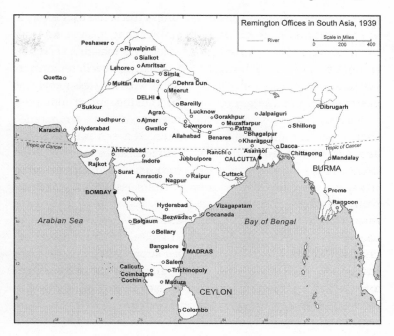

Remington Offices in South Asia, 1939

FIGURE 2.3. The distribution of Remington branches and outlets across South Asia in 1939 indicates how widespread the sale of typewriters had become. Map drawn by Julie Snook, FBCartS.

war sales spiked, American machines commanded 83 percent of imports in a market worth £363,314 (with a further £35,605 in spares). The United States maintained this supremacy through the Depression and into the late 1930s with over 90 percent of typewriter imports. As late as 1952, when less than a thousand British typewriters entered India and American domination of the market was beginning to slip, nearly two-thirds of imported machines were made in the United States.

That the typewriter market in India was dominated by American manufacturers is unsurprising given that it was "a predominantly American gadget."[34] Just as Singer ruled the sewing-machine world, so Remington, whose typewriter enterprise evolved from its small-arms expertise in the 1870s, held sway over India's typewriter empire. Occasionally, Underwood, the closest American competitor, protested against Remington's virtual monopoly in the supply of typewriters to government offices but to no avail. Like bicycles, portable typewriters (of which Remington claimed to have sold an astounding thirty-five thousand in India by 1928) were sufficiently cheap and desirable to be offered as prizes at sporting events or in promotions for other consumer items like cigarettes. Following a trajectory similar to Singer's a decade earlier, Remington steadily expanded its outlets in India. In 1920 the company had fifteen branches in India (as well as one in Colombo, Ceylon being another significant market for sewing machines, bicycles, and typewriters). By 1939, this had risen fourfold to sixty-two branches across India alone. As well as major administrative centers in British India, small but commercially vibrant towns, like Bezwada and Belgaum, had Remington offices. Other locations included the capitals of India's princely states, hill stations, rail junctions, cantonments, and regional hubs for the tea, indigo, and coal industries. Remington's Delhi head office controlled a further seven branches in Ceylon, Burma, Malaya, and Singapore, an illustration of India's centrality to the wider commerce of South and Southeast Asia.[35]

The Rise of the Rice Mill

The sample of everyday technologies considered in this book relate mainly to urban life, though, as we have seen, by the 1930s sewing machines and bicycles had infiltrated country areas as well. But, as pointed out previously, rural society was one of the principal locations, arguably *the* principal location, for colonial ideas of both technological inertia and "improvement." India was seen to be a largely rural, village-based society. As indigenous textile production shrank in the face of British industrial competition, the country's main contribution to overseas trade was seen to lie in its agricultural sector, with raw materials like jute and cotton, with such plantation commodities as tea and coffee, or foodstuffs like wheat and rice. Many of the trades that were seen to be most in need of technological innovation, or were thought destined for technological oblivion, were to be found in the villages, just as recurrent famine highlighted the vulnerability of weavers and other rural artisans hit by the loss of customary "entitlements." However, it is too simplistic to assume a stark distinction between urban and rural India when so many economic activities, including those of handloom weavers, spanned town and country or entailed a symbiotic relationship between the two.[36] The typewriter might remain an essentially urban phenomenon, but other everyday technologies like bicycles, sewing machines, and rice mills made use of close rural-urban linkages or brought small-time rural capitalists into close association with urban trade and industry.

As seen, too, in chapter 1, many of the exhibitions and fairs staged by the British in the second half of the nineteenth century were aimed at landlords and peasants. Much of the frustration felt by colonial officials at the apparent failure of these exercises in technological dissemination was rooted in the conviction that the countryside was so sunk in tradition as to be largely impermeable to technological change. The resistance of handloom weavers, many of them based in small towns and villages, to the introduction of the fly shuttle and other adaptations reinforced the perceived futility of expecting change in the countryside.

The rise of mechanized rice milling in the late nineteenth and early twentieth centuries demonstrates, by contrast, how small-scale mechanization could enter the countryside just as much as the towns. It further shows how radical, machine-led changes in the nature of food preparation and consumption had a social and technological impact comparable to the kind of changes that the sewing machine wrought with respect to clothing or the bicycle with regard to transport and urban employment.

Rice was one of the most widely consumed cereal crops in India as in much of Southeast and East Asia. But harvested rice grains (paddy) can't be eaten until the tough outer husk is removed. Traditionally, as discussed more fully in chapter 5, this was done by pounding the raw rice by hand until the broken husk could be separated from the inner grain. From the mid-nineteenth century, however, mechanical rice milling, using engines powered by steam, oil, or electricity, was introduced. The raw grain was propelled against a rapidly rotating drum or stone cylinder, or struck against a sharp blade (still known in south India as a "knife"), until the outer husk was struck off. Husked rice could then be subjected to further buffeting and polishing to produce a clean, white grain, from which even the inner skin (the pericarp) was removed.[37] Pioneered in the southern rice-growing regions of the United States, rice milling had begun to enter India, and more especially Burma (as part of the Indian Empire), with its huge, export-oriented commercial paddy production, in the closing decades of the nineteenth century. Unsurprisingly, given the origins of mechanical rice milling, the earliest machines employed were American Engelberg hullers, bringing India into indirect conversation with the plantation economies of the old South. Widely advertised and exhibited in early twentieth-century India, Engelberg machines remained in common use until at least the 1930s without achieving the market dominance associated with Singer sewing machines, Raleigh bicycles, and Remington typewriters. The component parts and overall design of the rice mill were much easier to replicate locally than the intricate mechanisms of the sewing machine or typewriter. Even before World War I rice mills were one of the machines most frequently encountered in paddy-growing regions. In Punjab, more a wheat-growing province than a rice-producing one, husking machines were replacing rice pounding in many districts by 1911.[38] Rice mills seldom arrived in technological isolation but in company with other small machines. By the 1940s, it was said to be a "common experience" in the Indian countryside "to hear the 'chug-chug' of a little engine either crushing sugar-cane, or pressing oil[seeds], or milling rice."[39]

It is, however, difficult to establish the scale of the importation of husking machines. Rice-milling equipment seldom appeared as a separate category in returns of India's seaborne trade, being subsumed instead under the amorphous heading of "machinery and mill work," along with flour mills, jute mills, and cotton gins. In 1920–21 machinery formed the third-largest category by value of imports into India, being worth an estimated £24 million (Rs 2.4 million). In 1929–30 rice- and flour-milling machinery were together worth Rs 2,374,691, but this slumped to Rs 881,091 in 1932–33 as the Depression deepened. By that time the bulk of milling machines came from Britain, followed by the United States and Germany; some mills were also being manufactured in India. A more dependable guide to the changing use and distribution of rice-mill technology is to be found in figures for the number of rice mills and their employees. These show the early lead and continuing dominance of Burma, until 1937 a province of British India, but also the gradual, if erratic, penetration of rice mills across the country.

Although rarely discussed in the same breath as other leading industrial enterprises, such as textile mills or railroads, by the early 1920s rice milling formed the sixth-largest industry in India in terms of the daily workforce employed, though at fifty-nine thousand workers this was only one-sixth the size of the cotton and jute industries. From the 1880s, up until about 1910, rice milling was heavily concentrated in southern Burma with a score or more large mills lining the banks of the Rangoon River, alongside the petrol storage tanks, sawmills, and timber wharfs that similarly typified Burma's export economy. In 1901 almost one in ten of the population of Rangoon was employed in rice mills. From about 1910 onward mills began to be established, on a far smaller scale than in Burma, in the paddy-growing areas of the Madras Presidency—principally along the canals and waterways of the Kistna-Godavari delta in Andhra and further south in the Cauvery River delta in Tanjore (Thanjavur). Unlike their counterparts in Burma, many of these "factories" consisted of little more than a single oil-powered engine and a solitary milling apparatus, under a thatch or corrugated iron roof and with crude brick or even mud-and-wattle walls. But others were grander affairs, with solid walls, a couple of oil or electric engines, and tall boiler-house chimneys. Like flour and sugar mills, cotton gins, cement works, and brickyard chimneys, rice mills announced the creeping industrialization of the countryside, a phenomenon evident to any traveler passing through rural India today. Between 1910 and 1930 rice mills proliferated rapidly in the Madras Presidency, erupting from towns like Ellore and Tanjore

TABLE 2.3. Rice mills in selected Indian provinces, 1900–1939

Year	Madras	Bengal	Bombay	Burma
1900	1	0	0	83
1901	1	1	0	96
1902	1	1	0	96
1903	2	1	0	108
1904	3	0	0	118
1905	4	0	0	128
1906	6	0	0	137
1907	23	0	0	153
1908	33	0	0	161
1909	48	0	0	164
1910	56	0	0	165
1911	62	0	0	165
1912*	68	0	0	245
1913	96	0	0	240
1914	110	0	0	258
1915	115	9	0	281
1916	127	35	0	311
1917	140	38	0	329
1918	142	40	0	332
1919	142	65	0	336
1920	149	88	0	353
1921	157	91	0	429
1922	278	101	38	482
1923	357	117	56	529
1924	350	132	61	518
1925	355	191	70	543
1926	391	235	65	567
1927	444	277	65	572
1928	462	286	73	606
1929	459	312	82	608
1930	463	315	80	613
1931	317	314	77	589

TABLE 2.3. (continued)

1932	302	330	88	609
1933	313	337	84	618
1934	332	343	84	637
1935*	328	346	86	647
1936	326	381	4**	657
1937	326	394	3	650
1938	342	411	3	683
1939	341	400	4	692

Source: Annual reports on the Indian Factory Act by province.
* New legislation in force from 1912 and 1935 changed the definition of a factory according to the size of its workforce.
** Following the separation of Sind from Bombay, 1936.

TABLE 2.4. Average daily number of rice-mill workers, 1910–1940

Year	Madras	Bengal	Bombay	Punjab	Burma
1910	n/a	0	0	0	18,478
1920	8,089	3,604	n/a	186	36,486
1930	15,796	12,225	777	429	42,137
1940	12,783	18,447	163*	366	41,626

Source: Annual reports on the Indian Factory Act by province.
*The apparent decline in the Bombay Presidency was due to the creation of a separate province of Sind in 1936.

into surrounding villages. The great majority of south Indian mills were owned by local entrepreneurs, members of "rich peasant" and trading communities for whom the move into rice milling represented an opportunistic shift away from unprofitable agriculture into the more profitable realm of factory ownership, food processing, and small-scale manufacturing. Across India by 1937–38 there were estimated to be 1,135 rice mills operating throughout the year, with a further 175 functioning on a seasonal basis with the local rice harvest. Perennial mills employed 43,579 workers (with a further 3,710 workers in seasonal factories). With more than 400 each, the provinces of Madras and Bengal held the largest number of mills, but between 20 and 70 mills were also to be found in Bihar, Orissa, and Assam. In Punjab there had been hardly any rice mills before 1918, but by 1924 there were 25 of them, employing 571 workers.[40]

Rice milling was one of a cluster of expanding rural industries that used modern machines to process agricultural commodities: these included flour milling, cotton ginning, the crushing of sugarcane and oilseeds, and the decortication of groundnuts. These emerging agrarian industries were linked in various ways—through their dependence on global as well as local markets, in their shared use of power-driven machinery, and in the largely unskilled workforce they employed. A single factory might serve several purposes, moving from one commodity to another according to the harvest season or the strength of demand. But the milling industry was also unstable. In the Madras Presidency the hunt for quick profits caused the industry to become overextended in the 1920s, and it struggled to compete with cheap rice from Burma. As the Depression hit, the number of working mills in Madras slumped, as it did in other provinces, falling from 463 in 1930 to 317 a year later. The number of mills and their employees only gradually recovered and even in 1941 stood well below the 1930 figure.

Unlike other everyday technologies considered here, in which the more conventional power of hand and foot was primary, the rise of the rice-milling industry in India was heavily dependent upon the availability of steam power and still more the power generated by oil and electric engines. That reliance depended in turn on wider market conditions and a degree of state support. Madras was in the forefront here, encouraging the early spread of steam- and oil-powered engines and by the mid-1930s low-cost electricity supplied from government hydroelectric schemes. Chatterton, whom we have already encountered as a proponent of state-aided industrial development in early twentieth-century Madras, noted with some pride in 1911 that, in the main paddy-growing districts of Kistna and Godavari and further south in

Tanjore, "the primitive methods of husking by hand have to a large extent been superseded by modern machinery." Partly through state assistance but also as a result of "progressive private effort," Chatterton commented, "a number of what may be termed rural factories have come into existence, which use machine processes usually on the smallest scale that it is practicable to employ them." Given the high cost of other fuels in Madras, "the internal combustion engine, on account of its very high efficiency, especially in engines of small power, is already very largely employed, and is likely to become in time almost the sole source of power." Here, for Chatterton at least, was evidence that, given the right environment and appropriate machinery, "improvement" could work.[41] However, in 1910, acting in the name of laissez-faire orthodoxy, the secretary of state for India, John Morley, reined back state support for industry, and it was again left to private enterprise to foster the growth of rice milling and other rural and small-town enterprises.[42]

× × × × × × × × × × × × × ×

Sewing machines, bicycles, typewriters, and rice mills were core elements in a wide-ranging process of technological diffusion and adaptation in India from the 1880s onward. These particular technologies and the goods, skills, and services associated with them were not the only ones that could be selected to exemplify this development, but they were among the most visible and momentous. Together they spanned town and countryside, encompassed princely states and British India, and marked out a role for the innovative (or greatly extended) mechanisms of pedal, treadle, keyboard, and rotating mill that were subsequently applied to other mechanical devices. The success of these modern machines was tied to their versatility, their low running costs, the small capital investment needed for their acquisition, the minimal levels of skill needed for their use, and the limited degree of state funding and sponsorship required. Although the supply of such machines was constrained by often adverse economic conditions and by the fluctuations of international trade, there was a generally upward trajectory to their importation and use into the 1960s, by which time indigenous production largely replaced foreign-made goods. As examples like the sewing machine and bicycle suggest, the scale of use and the impact on society was less pronounced in India than in much of Europe and North America. Yet even so, the advent of these new technologies was beginning to have a substantial impact on the way Indians lived and worked, and to serve as evidence of a far wider Indian appetite for technological change.

CHAPTER THREE

Technology, Race, and Gender

Except in connection with slavery and the racial prescriptions that governed technological innovation (or its absence) in slave-holding societies, the history of technology has rarely engaged with issues of race.[1] By contrast with the history of bodies, which in recent scholarship have been comprehensively sexed, classed, and ethnicized, machines have been widely discussed in connection with gender and class but seldom in association with race and ethnicity.[2] And yet, throughout most of the period from the 1880s to the 1940s, race had a very significant bearing on the understanding and use of technology, especially everyday technology, in India as in many other European colonies. Who sold a machine to whom, who used that technology or was displaced by it, who gained prestige or lost status by its use—these were questions in which issues of race repeatedly asserted themselves.

Just as race can be seen as "the most obvious mark of colonial difference," so can it be understood as one of the main ways in which technology acquired social meaning.[3] Race was used to signify technological aptitude (or the lack of it), and to place the stamp of seemingly scientific authority on claims for modernity or for its denial. And yet race seldom operated alone. It was commonly qualified or reinforced by reference to class and gender, both of which had a critical role in defining the uses to which technology was put and the meanings assigned to it. But, in India, as in many other colonized societies, race ideology and practice tended to take precedence over class or subsumed issues of gender. There was, though, always an element of instability in the racialization, as in the gendering, of technology. And there always lurked the subversive possibility of empirical and ideological challenges to ascribed race and gender attributes. Our interest in this chapter thus lies with issues of how, by whom, and to whom small machines were sold or by what other routes they encapsulated, transgressed, or transformed racial boundaries and gender divisions.

Sewing Machines as Racial Goods

In the West, guided by considerations of gender, the bulk of sewing machines were sold for women's use, especially in the home. Sewing machines became closely associated with women's work and the rise of women consumers.[4] Singer prided itself on selling machines that did more than perform a utilitarian task. Especially after the introduction of its "new family model" in 1865, its machines were intended to be attractive pieces of furniture that would sit stylishly in any middle-class parlor. The domestic context of the sewing machine was further exemplified by the way in which in Western countries the salesman, or his wife, would visit the purchaser at home to explain how her gleaming new machine worked.[5]

Initially, Singer, which sought to pursue a common sales strategy throughout its vast commercial empire, thought in similar terms with respect to the Indian market.[6] There, however, with gender transmuted by race, the assumption was that only women who were either Europeans or Eurasians (officially designated "Anglo-Indians" from 1911 onward) were likely to buy and use such innovative machines. Europeans (and the *darzis* they employed) were certainly among the first to use sewing machines in India. They either purchased them from local agents or brought them in their household effects from Britain. Others were sold or passed on secondhand among European residents. It was widely believed (as we saw in chapter 1) that among Indians the "stitch-less" nature of their clothing, especially women's clothing, made the sewing machine redundant. It was further assumed that Indians in general, and Indian women in particular, were incapable of using such machines or even comprehending their use. In 1888 John Mitchell, one of the "traveling examiners" Singer periodically dispatched from New York or London to

report on local sales, remarked that Indian women were so ignorant and secluded that "if advertisements reached their hands
they would be unable to understand them." His brief tour of India had convinced him of the "absolute uselessness of the Sewing
Machine for the vast majority of the population," whose dress
appeared to consist of "one or two plain pieces of cloth wound
in curious folds around the figure."[7]

It followed, to cite another Singer representative in 1884, that
the "vast native domestic population is closed to us as yet and
will be so till western ideas and western dress take a firmer hold
on the masses."[8] Pursuing this racial logic Mitchell went through
the 1881 census to establish how many Europeans and Eurasians
were resident in particular towns, districts, and cantonments,
and hence how many machines Singer could hope to sell. Given
the smallness of the white population in India (and that households rarely ever purchased more than one sewing machine) it
was not likely to be many: in the 1880s Singer struggled to sell
more than two thousand machines a year. And yet, in the confi-
dent spirit of international capitalism, Singer liked to represent
its machines as bringing civilization to the world. A company
trade card issued in 1892, in time for the Chicago World Fair,
claimed that for twenty years in India Singer had "been a factor in helping the people of India toward a better civilization."[9]
But privately Singer representatives doubted that the sewing
machine had much of a future there until civilization, as they
understood it, gained a more secure foothold.

There was, though, an alternative strategy. In 1875, shortly after it had begun operations in India, Singer acquired as its agent
in Bombay an enterprising young Parsi, Nasarvanji Mervanji
Patell, whose career as the company's principal agent in South
Asia lasted until his retirement in 1911. Patell belonged to one
of the leading Parsi families in Bombay and hence to the community that stood at the forefront of the city's commercial and
industrial development.[10] As entrepreneurs and technological
intermediaries between Europeans and the mass of the population, Parsis occupied a crucial role in the dissemination and
popularization of everyday technology in India. In Bombay and

in its hinterland from the Deccan in the south to Punjab in the north, and as far east as Allahabad, Parsi stores sold bicycles, motorcycles, typewriters, sewing machines, cameras, gramophones, and other "mechanical novelties." Patell's family was at first displeased by his decision to sell sewing machines, which they considered a "common shop business."[11] But he persisted and began to devise a new marketing strategy for a company whose sales in the late 1870s were behind those of its main American rival, Wheeler and Wilson.

Patell's strategy was to bypass the small number of European purchasers and focus instead on Indian customers. He particularly sought to target Indian tailors, who potentially constituted a far larger market, and the community or caste leaders who (as master tailors themselves) made the critical decisions about purchasing or hiring sewing machines. His strategy was not easy to implement. He had to fight off or take over rival agencies and contend, through the law courts if necessary, with traders who sold "counterfeit" Singers (mostly made by rival European exporters). Patell took the view that, while European customers were primarily interested in the appearance of the machine and then in how efficiently it worked, Indians "as a rule look to cheapness." They would buy less expensive machines, including "imitation Singers," unless Singer sold its machines at a competitive price and was prepared, contrary to company policy, to offer discounts of at least 10 percent.[12] Following practices common elsewhere but still new to India, Patell and his agents drew up hire-purchase and loan agreements that enabled tailors who could not afford to buy a machine (costing at the time between Rs 50 and Rs 100) to rent or own one on easy terms. As a result Singer soon disposed of more machines by hire purchase than were sold outright. In 1877, for instance, 1,093 machines were sold (599 of them to Indian customers), but a further 1,508 went through hire purchase (885 to Indians).[13]

Patell sought to persuade the principal tailors in Bombay and Surat, known to him as the "Chanchias," through their headman, Bacher Ghella, to abandon the Wheeler and Wilson machines with which they had been familiar since the 1850s and adopt

FIGURE 3.1. N. M. Patell, Singer's principal agent in India in his office. Undated photograph, unknown photographer, reproduced from the Singer Archive by kind permission of the Wisconsin Historical Society, Madison, WI (WHi-90488).

Singers instead. It is not entirely clear who the Chanchias were, but it is likely that they were a Gujarati caste more normally associated with the sale of domestic cooking oil, though in south Gujarat the term *Ghanchi-Gola* was applied collectively to several low-ranking artisan and trading communities. Winning them over to Singers involved a protracted struggle. The Chanchias found the more open structure of the Wheeler and Wilson machines better suited to the light but bulky cotton shirting they used and their rapid manner of work. It took Patell more than twenty years to overcome their resistance—not to sewing machines, which the Chanchias had readily adopted, but to the unfamiliar Singer machines.

Selling Singers

Patell's career demonstrates the racial terms in which Singer, like many other international firms, saw itself conducting business in India. But it also shows the critical role that Indian agency could have in selling global goods in a local market. The fact that India was seen to be a difficult place to sell sewing machines gave Patell a degree of freedom that he would not have had in most other countries.[14] Despite the relatively small number of machines sold in India, Patell's strategy was, in local terms, highly successful. Singer sales rose steadily, and, having early on established agents in Lahore and Allahabad, the number of branches under his control increased almost yearly. In 1885 there were about thirty-five Singer outlets. By 1905 the company had more than a hundred branches across India, Ceylon, and Burma.[15] Increasingly its shops were located in "native towns" and bazaars rather than European shopping areas. Even small towns could boast a Singer shop, and company representatives roamed the countryside in search of new customers.

But while Patell was largely the architect of local Singer sales success, he was not without his critics and detractors—and here, too, race played a part. The visiting inspectors sent from London to assess the progress of Singer operations in India in the 1880s took a skeptical view of Patell and his business methods. They found the stock in his showrooms dirty, rusty, and poorly displayed. They questioned the bookkeeping practices of Patell and his office manager, who just happened to be his brother-in-law. They criticized the "narrowness" of his marketing ideas, his failure to follow the "broad principles" of the company sales policy, and his apparent lack of interest in selling to the customers they thought really mattered—Europeans. They attacked Patell himself, finding him fussy, driven by "personal spite" against those he considered his enemies, obsessed with trying to sell their machines as cheaply as possible, while appearing offhand and brusque with his customers. Even if they accepted that Patell was "energetic and earnest," they attacked him as a Parsi, as a virtual foreigner (the Parsis having migrated from Persia centuries earlier), as a member of a community who, outside Bombay, was as remote from the languages, customs, and manners of the local inhabitants as Europeans themselves. In other words, Patell's "race" was used against him and the business methods he employed to sell sewing machines.[16] It should be noted in passing, though, that this hostile attitude was at variance with the more common view of the Parsis as one of India's most "westernized" communities, with a "genius for business."[17] In June 1887 Patell was forced to resign, only to be promptly reinstated. Crucial to his survival was the support of Singer's managing director, George R. McKenzie, with whom, through their frequent correspondence, Patell had built up a close working relationship.

Patell in turn warned McKenzie about the ways in which Europeans treated Indians like himself. "Nine hundred and ninety-nine out of a thousand Europeans who come out to India," he confided as he waited anxiously for the arrival of London's latest emissary, D. Davidson, "change their attitude, manners, and politeness and try to be Lords in India." Hence, "Mr Davidson might in time feel my suggestion[s] awkward and uncalled for."[18] Events confirmed Patell's worst fears, and yet, despite pleading his local expertise and knowledge of "the habits of the people here," he needed to make compromises to survive. He agreed to sack his brother-in-law, improve the bookkeeping, and clean up the showrooms, but he also made other concessions that ran counter to his own inclinations. He dismissed as impractical the suggestion that every branch office should be run by a "smart European," but, persisting in the belief that Europeans were the main market, Davidson obliged Patell to hire three Eurasians as itinerant canvassers, though Patell had "very little confidence in these East Indian youths." He even had to accept Eurasian or European women as shop assistants in the Bombay, Calcutta, and Madras showrooms to facilitate sales to white customers.[19]

Just as the Parsis were one type of technological intermediary in the racial and communal configuration of colonial India, so Eurasians were another. By virtue of their mixed racial origins and Western lifestyles, Eurasians were assumed to be more European than Indian in their aptitude for European technology while also having the advantage of (some) familiarity with Indian languages and customs. Cheaper to employ than Europeans, they were given privileged status on the railroads, in the telegraph service, the police, the medical service, and other state agencies. As shop assistants, Eurasians were particularly expected to attend to European customers, but Eurasian men also worked as chauffeurs, clerks, and overseers, and as technicians in factories, workshops, and plantations. Singer's London agents were therefore following an established pattern in believing that Eurasians were the key to selling sewing machines, especially since they (like their counterparts in the United States and Britain) could visit Europeans at home in a way that Indians generally could not.

Patell did not actively sabotage this racial strategy, but he was hardly sorry when it went awry. Of the three Eurasian salesmen, Doyle died at Trichinopoly (but not before he had misappropriated Rs 83), and Allen was jailed for absconding with Singer funds. Worst of all was the third, Arnold, of whom Patell declared, "I have never found such a daring rogue." He failed to keep proper accounts, sold only two machines, and was jailed for cheating Singer of more than Rs 500.[20] While the Eurasian experiment failed, Patell was forced to recognize that a more aggressive sales policy was required. But, in the main, he was allowed a degree of latitude, rare among Singer employees, to decide how best to sell sewing machines in his own country.

Technology Engendered

In 1892, when Singer produced a series of colorful trade cards showing customers around the world in traditional dress, two separate designs were produced for India. One showed Patell and his Bombay office staff with a man seated at a treadle sewing machine. The other depicted a *darzi* and his wife. Unlike the images used for many other countries, where a woman was shown working the machine, on the Indian card she was depicted sitting on a chair alongside the machine and next to her standing husband as if in a kind of technological limbo. It looked as though they were both uncertain who should use the machine.[21] In shifting the focus of Singer sales from Europeans to Indians, Patell helped, however inadvertently, to effect a gender change as well. Whether as customers or sales assistants, women were not central to his marketing strategy. For him selling sewing machines in India was essentially about selling them to men, particularly to men who by custom and trade were tailors (as was also the case, at least initially, in many other regions, including the Middle East).[22] Nonetheless, it is clear that an increasing number of women in India, including in the Parsi community, were using sewing machines or learning to sew on such machines by the early twentieth century.[23] How and why did this gender shift occur?

In part the answer returns us to questions of race, as well as of class and gender. From early in the colonial period sewing, along with embroidery, dressmaking, and millinery, was seen as a suitable occupation for poor white or mixed-race female orphans (as in the military orphanages of Calcutta) or among the destitute and "fallen" women taken up by Christian charities. The female inmates or beneficiaries of such institutions and organizations were expected to sew—in part to earn money for themselves and the charities that supported them and in order to prepare themselves for future lives as housewives and seamstresses, but also because the discipline of sewing was seen to be conducive to feminine patience, diligence, and appropriate moral conduct. However, sewing also transcended the racial divide. Making and mending family clothes was a part-time occupation for many Indian women and was considered a useful and respectable form of women's home work. By extension, as in many other parts of the world, sewing and the use of the sewing machine came to be seen as a desirable part of a modern woman's upbringing

FIGURE 3.2. One of the many trade cards produced by Singer c. 1892 to show the worldwide distribution of its sewing machines and the colorful costumes of their users. Here an Indian *darzi* and his wife pose alongside a Singer treadle machine. Reproduced, with permission, from the Warshaw Collection of Business Americana—Sewing Machines, Archives Center, National Museum of American History, Smithsonian Institution, Washington, DC.

and domestic accomplishments. As Kupferschmidt observes, "In societies . . . which entertain strong reservations against women working or trading outside the home, the sewing machine comes as a perfect solution."[24] But the sewing machine also became part of the repertoire of Indian women's uplift and social reform in the second half of the nineteenth century and in the early twentieth century. In literature produced by Hindu reform organizations like the Arya Samaj or the Sikhs' Singh Sabha women were urged to take up useful domestic tasks like sewing and to cut costs or augment family incomes by making their own clothes. In one reformist tract of the period, the dutiful housewife not only runs her own home efficiently but also helps a neighboring widow to improve her domestic finances by investing in an old sewing machine.[25]

To take a more extreme example, when women from the "criminal tribe" of the Bhantus (or Sansis) from northern India were sent to the penal settlement of Port Blair in the Andaman Islands in the 1920s they were set to work sewing; some were introduced to the use of sewing machines. In this case, the use of sewing and sewing machines as an instrument of reform was all the more marked since the Sansis were renowned for being scantily dressed "vagrants" and "nomads," barely inducted into "domiciliary civilization" and still "in the suckling stage of human progress."[26] The sewing machine became a means of fixing them in one place: it obliged them to undertake "useful labour" and submit to social and moral reform. Sewing and embroidery were similarly disciplinary tasks imposed upon young women in "criminal settlements" located in mainland India.[27]

The rehabilitation of criminals was not the only route by which women's use of the sewing machine spread. When self-help associations for women were set up in Bengal from 1913 onward by Saroj Nalini Dutt, sewing was among the activities women were encouraged to adopt for their gainful employment. Some acquired sewing machines or were given them by well-wishers.[28] One of the ways in which the many thousands of women who were displaced and became refugees as a result of partition in 1947 were given work was by providing them with

FIGURE 3.3. Bhantu women from north India pictured with sewing machines at Port Blair penal settlement in the Andamans. Reproduced from the Ferrar Collection, album 1, by kind permission of Dr. Kevin Greenbank, Centre of South Asian Studies, Cambridge, UK.

sewing machines.[29] After Indian independence state-run rural development and welfare schemes typically expected women to sew, embroider, and ply sewing machines; men, by contrast, were assigned to outdoor construction work and agricultural tasks.[30] But the use of sewing machines crossed class as well as racial and gender boundaries. Among the Indian middle classes it appears by the interwar years to have been fairly widespread. Madhur Jaffrey describes how in her family's high-status household in Delhi in the 1930s her mother had a sewing machine on which she made clothing for the family while employing a *darzi* to "do the simple stuff"—routine items of dress that required little artistry or imagination.[31]

Across the social spectrum the sewing machine helped to define but also to enlarge the sphere of women's work. By the mid twentieth century the machine had become one of the most widely disseminated domestic appliances in India, one of the few machines made readily available to women and considered suitable for their use. In a society where machines more commonly

denoted men's work rather than women's, the sewing machine
enabled women to exploit the practical advantages and prestige
of machine use and to earn money while remaining within the
respectability of the home.[32] In this regard, the domestic history
of the sewing machine contradicts the idea that the Indian home
remained a strictly segregated space, untouched by the exterior
world of science, technology, and consumption.[33]

Singer and other manufacturers began early in the twentieth
century to appeal directly to middle-class Indian women and,
as in the West, to present sewing machines as highly desirable
domestic goods. Advertisements appeared in newspapers and
journals showing smiling, sari-wearing women, working happily
at their machines. One attractive lithograph image produced by
Singer around 1900 or even earlier showed an Indian woman at
work on her machine, closely watched by a young girl, whom
one assumes to be her daughter (see frontispiece). As part of
their emerging identity as instruments of women's work, sewing
machines became objects men gave to women—for instance as
philanthropic gifts to women's associations and welfare societies or as prizes for schoolgirls and female college students. Like
bicycles, and later radios, motor scooters, and automobiles, sewing machines became highly prized dowry presents, given by the
bride's family to the groom's. But where bicycles were likely to be
used by the male members of the receiving household, the gift of
a sewing machine emphasized the new wife's domestic responsibilities. Dowry gifting was one of the most important social
mechanisms by which new consumer goods were disseminated
among middle-class households in India. Indeed, at a time when
social reformers were calling for the "evil practice" of dowry giving to be abolished, the spread of sewing machines and bicycles
through the social mechanism of the dowry demonstrated the
increasing acquisitiveness of the rising middle class.[34]

But, as emblems of women's work, sewing machines also acquired—at least in the minds of men—the negative associations
of a burdensome domesticity. In Bibhutibhushan Bandyopadhyaya's novel *Aranyak: Of the Forest*, published in 1937, the narrator
nostalgically recalls the years spent in his youth among the trib-

als and forests of rural Bhagalpur, but this serves to heighten the contrast with his present situation as a responsible family man in Calcutta. "I have left behind my life of freedom [in the jungle] and have become a householder," he laments. "I sit in my room in a narrow little lane in Calcutta and listen to my wife at her sewing machine."[35] As with many other everyday technologies that announced themselves through their audibility, the repetitive sound of the sewing machine highlights for the male narrator the contrast between youth and experience, freedom and domesticity.

Bicycle Races and Gendered Cycles

In the 1890s and 1900s bicycles established themselves in India as a popular mode of recreation and personal transport for Europeans, as they did for white populations in other parts of colonial Asia.[36] The bicycle gave Europeans a newfound sense of physical freedom—to roam around towns, hill stations, and cantonments, to venture out into the countryside, to tour archaeological sites, or to visit the encampments that sprang up for the Delhi Durbar of 1911. In a new sociability, following a trend already established in Europe, in many towns and hill stations across India white cycling clubs sprang up. When Ronald Ross, then a junior medical officer, was stationed at Bangalore (Bengaluru) between 1895 and 1897, he and his wife belonged to the local cycling club. At Secunderabad, his next posting, Ross cycled around town, musing on the "malaria problem" while riding his "excellent bicycle," rather than using a conventional pony and trap.[37] A bicycle was substantially cheaper to buy and maintain than a horse and so was available even to low-ranking Europeans, for whom the cost of a horse and *syce* (groom) was prohibitive. Before 1914 bicycles were also used as a means of transport by relatively high-ranking Europeans, like police superintendents and factory inspectors, but this phase of the bicycle as a high-status European commodity passed fairly rapidly—certainly by 1920. When Gilbert Slater arrived in 1915 to take up a chair in economics at Madras University, it was considered exceptional (and rather improper) for a

European of his standing to cycle around the city or ride off into the countryside. He soon switched to a motorcycle. Rather like the Japanese housewives whom Robin LeBlanc characterizes as "bicycle citizens," for many Europeans riding a bicycle became a means of identifying themselves as being on the margins of the white elite.[38] Cycling in the heat of the Indian day, a European was likely to arrive at his or her destination sweaty and disheveled, his or her racial authority compromised. A horse, a motorcycle, better still an automobile, was a more dignified proposition and preserved greater social distance between colonizer and colonized.

Europeans began to feel vulnerable on bicycles—from fear of being jostled in the bazaar or attacked during a riot. A critical moment came in April 1919, when Miss Sherwood, an English missionary doctor, riding alone through the backstreets of Amritsar in Punjab was knocked off her bicycle and beaten up by a crowd of nationalist agitators.[39] The treatment she received became a rationale for the imposition of draconian measures across the city, including the confiscation of all Indian-owned bicycles. But the relationship between bicycles and European women was not unambiguous. During World War II, when automobiles were scarce and petrol rationed, some white women took to riding old and battered bicycles as the only means of transport available to them. In Paul Scott's end-of-the-Raj novel, *The Jewel in the Crown*, the missionary Edwina Crane cycles around Mayapore on a "ramshackle-looking but sturdy Raleigh." If she remains a memsahib in the eyes of her Indian assistant, it is "in spite of the bicycle."[40] Daphne Manners, a volunteer nurse, also rides a bicycle and uses it to meet her Indian lover Hari Kumar. Their bicycles become a symbol of racial and sexual transgression, especially when Manners is raped, and their abandoned cycles become critical evidence for the case Ronald Merrick, the investigating police officer, fabricates to entrap Kumar.

Given its affordability and limited status value, the bicycle was never a European monopoly. In Calcutta and Bombay, Indians set up and joined cycling clubs as early as the 1890s. Young, middle-class men, particularly students, took to the bicycle, went on cycle tours in the countryside, and participated in cycle races. In Bengal the bicycle acquired the kudos of being healthy—for the race, and not just the individual. Even though the machines remained foreign, or were at best assembled locally from foreign parts, popularizing their use became a means of countering, on however modest a scale, European representations of Bengali men as "effeminate" and "effete." Winning a cycle race might not transform Bengalis into a "martial race" overnight, but it con-

tributed to a wider, emerging cult of physical fitness. Touted as physically and morally healthy, cycling helped promote an image of Indian energy and self-reliance.[41] As well as bicycle races, which had become major sporting events by the 1910s, in the interwar years Indians embarked on cycle tours across India and beyond. A young Parsi, Scouter Davar, set off in January 1924 on a solo journey to "encircle the world" that lasted seven years. Having covered sixty-five thousand miles, crossed the Sahara, penetrated African jungles, and scaled Andean passes, he returned home to a hero's welcome.[42] That several leading cycle "tourists" and racers were Parsis is indicative not only of Parsi "anglophilism" but also of growing concern within such a small community that, if their reproduction rate continued to dwindle, they would face collective extinction. Cycling and other active sports was one way of trying to maintain their physical vigor.[43]

At what point—and to what extent—Indian women became cyclists is less clear. In Europe and North America cycling has been associated with "the new woman," and with growing social, even sexual, freedom for women.[44] This was less evidently the case in India. It is not difficult to find examples of women from elite families who rode bicycles, at least in their youth. Lado Rani Zutshi and her daughter Manmohini Sahgal, members of the extended Nehru family of Kashmiri Brahmins, rode bicycles and went horse riding, but, in the early twentieth century, these were activities "rarely engaged in by women."[45] In Rawalpindi in the 1940s it was rare for girls to ride bikes in public, even to go to school, while in Lahore men from the countryside gawped in amazement at the unfamiliar sight of women on bicycles.[46] In general cycling was not common once women reached marriageable age. There was an argument (made mostly by men) that it was dangerous for young women to ride bicycles for fear of rupturing their hymens and so ruining their marriage prospects.

There seems to have been a broad prejudice against the physical mobility and independence women might acquire by riding a bicycle. In rural Maharashtra as late as the 1970s Hemlata Dandekar noted how "you never see women or even young girls on bicycles in the village." Even in the nearby town, "a woman riding a bicycle is a rare sight." One girl told her that she wouldn't dare ride a bicycle in her village—her in-laws would be scandalized. Even those who rode bikes as girls were prevented from doing so as adults: it was considered no more appropriate for a woman to ride a bike than to plow a field or drive a tractor, conventionally men's work. Some of the women Dandekar interviewed clearly resented this restriction.[47] So, while for young men the bicycle might be a means to adventure (including visits to cinemas and brothels), for women the bicycle symbolized the constraints patriarchy imposed on their adult lives.

But some women *did* ride bicycles and by the 1940s some advertising was targeted specifically at them. Others knew how to cycle when the occasion required it. Several of the accounts given by women of their involvement in the communist-led Telengana movement in Hyderabad in the late 1940s refer to their use of bicycles. Women typically rode bicycles to act as couriers for the rebels or to help insurgents elude police operations. Moturi Udayam recounted how, when the police came looking for a political activist hiding in her house, she claimed that the bicycle standing outside was not his but her own. "'How can it be your cycle?' the policeman asked, 'as if women ride cycles.' And I said, 'Why not?'" To prove her point, she got on the bike and rode round the house several times. But Moturi presented herself as an exceptionally feisty woman who did what she liked. "I have no superstitions," she told her interviewers. "I have achieved whatever I wanted. From learning to cycle to [doing] my job."[48] An association between cycling and women's status was not just historically important: the riding of bicycles and motor scooters in public remains a salient indicator of women's independence in India today.[49]

Caste, Race, and Typewriters

Like government offices, commercial firms and private employers first took to the use of typewriters in India in the closing years of the nineteenth century. Patell in Bombay began to use a typewriter (or, more likely, to have his letters typed for him) in 1895. As was frequently the case with the introduction of what rapidly became everyday technologies, the transition from the handwritten to the typed letter occasioned little comment. However, the rapid adoption of the typewriter and the growing need for typists can be traced through the increasing number of advertisements in the wanted columns of English-language daily newspapers like the *Bombay Chronicle* or the Calcutta *Statesman* for typists and stenographers. In January 1922, for instance, advertisers in the *Statesman* sought an experienced Anglo-Indian male typist for a European firm at Mirzapur, a shorthand typist for an up-country tea estate, and a "competent lady shorthand typist" for a newspaper office in Calcutta, while Berry & Co., seeking an experienced shorthand typist, promised a "good salary to a fast and accurate man." Among those offering themselves for employment were a Madras man with three years' typing experience, a female shorthand typist who had previously worked for a European firm, and an "Experienced Lady" who sought to work from home as a shorthand typist ("own typewriter, neat and accurate").[50]

India had a long and sophisticated scribal tradition. Developed under the bureaucratic empires of the Mughals and Marathas and at a series of regional courts, this reservoir of clerical skill was readily utilized by the incoming British. It was partly from among traditional "service communities," such as Brahmins and Kayasthas, that the clerks, typists, and stenographers of the modern government and business office were recruited. It is no surprise to find among the list of clerks in the Government of India's home department in the early twentieth century a sizeable proportion of Tamil and Bengali Brahmins. In one instance, five out of six applicants for posts as shorthand typists were Tamil Brahmins.[51] In this respect, as in many other areas of society affected by technological change, communities retooled their old skills to meet current needs.

But, if in some parts of the world the typewriter symbolized an exciting new world of reliability, speed, and individual self-expression, in India it seemed, at least at times, to suggest the more negative side of modernity. The clerks (*kerani*) described by Sumit Sarkar for late nineteenth- and early twentieth-century Calcutta came from the poorer sections of the Bengali *bhadralok* (the "respectable" castes of Brahmins, Vaidyas, and Kayasthas). These were individuals whose literacy and high-caste status appeared barely rewarded by the tedium of their working lives, the paltry nature of their salaries, and the racial abuse they received in government offices and merchant houses around the city. As Sarkar puts it, by the late nineteenth century office work and the clock time that governed it "had come to signify all that was demeaning and oppressive in colonial bhadralok life." Or, as another recent historian puts it, the colonial clerk suffered a life of "deprivation and domination."[52] An equally negative impression is conveyed by the Bengali writer Bibhutibhushan Bandyopadhyaya. In his novel *Aparajito*, Nripen, a friend of Apu, the book's central figure, toils in an office till seven at night, enduring a "small, closed, dark, claustrophobic life." For Apu, along with congested streets and noisy traffic, the office and its clamorous typewriters represent the tyranny of "all things modern" and leave him craving the quiet, open spaces of the Bengali countryside.[53] Nationalist critics, too, railed against a colonial order which reduced educated men to mere clerks when, under a more enlightened regime, they might have found recognition as entrepreneurs and intellectuals. And yet, for all this, the typewriter and the modern office did help to institute change. Colonial rule increased the availability of literacy and education, and favored the extension of technical skills like typing to communities that had previously been excluded from bureaucratic life, such as Christians, low-caste Hindus, and Sikhs. This began to erode the monopoly of the old scribal classes, a process further aided in state service by a system of communal representation.

Proficient use of a typewriter might call for some degree of formal instruction—though it is clear that in India, as elsewhere, many individuals learned to type by themselves, perhaps with help from a family member or friend. Acquiring a job, especially in a government office, was usually dependent on having a recognized qualification or a certificate of competence from one of the many typing and secretarial schools that sprang up in India from the 1890s onward, especially in the commercial districts of Bombay, Calcutta, and Madras. By the 1920s Bombay had several such schools. The Kalbadevi Shorthand and Typewriting Institution, dating from 1898, was eager to attract clients and offered "special arrangements for ladies." The nearby Popular Shorthand and Typewriting Institution also gave concessions to women, the poor, and the unemployed.[54] Many such schools prepared pupils for international shorthand and typing examinations, such as Pitman's. In a model shared by other everyday enterprises (such as "Usha" sewing machines after independence), manufacturers and importers organized courses to teach typing and encourage the use of their own machines. Remington, which established its own schools in India to "train the Babu or educated native in the 'twin arts' [of typing and stenography]," produced the majority of typists entering government service.[55]

Clerks with typing and shorthand qualifications received a higher salary than those without them, earning by the early 1930s between Rs 60 and Rs 75 a month, enough to place a typist among better-off office workers. But this did not lift the mass of the clerical underclass out of poverty or free them from the prevailing sense of oppression, discrimination, and hopelessness that Sarkar describes. An article in the *Social Service Quarterly* in 1926 depicted the miserable existence of lower middle-class men in Bombay who earned between Rs 50 and Rs 250 a month. Trudging daily to and from work (commuting by tram or train or walking to save money), they had to endure long office hours, while their wives remained at home, barely able to meet food bills and housing costs. Their lives were "neither . . . comfortable nor absolutely destitute."[56] In these circumstances it is not surprising that sections of the clerical workforce sought to organize themselves in order to better their conditions and communicate their grievances. As early as 1893 a Shorthand Writers' Association was established in Madras. In return for an annual subscription of Rs 5, it promised to help members find employment as shorthand writers and typists in both the city and up-country. Forty years on, the Stenographers' Association in Madras was set up for a similar purpose. But, despite poor work prospects, by the 1930s almost every middle-sized town in India had its own secretarial school, technical college, and commercial institute, churning out still more typists, bookkeepers, and clerks.

Race and Gender in the Office

The typewriter could serve to project the image of the heroic male, especially, in colonial times, the heroic white male. An advertisement for Remington portables in the 1930s showed a European official, presumably on tour, working outdoors at his desk, while a liveried Indian servant stands by, clearly proud of his employer's ownership of so splendid a machine.[57] Almost thirty years earlier Captain Gillespie of the Royal Engineers wrote enthusiastically of the American-made Blickensderfer portable he had taken with him on an expedition to Tibet: "It accompanied me on the march, travelling on a coolie's back. I used it under all sorts of conditions, in pouring rain, at over 13,000 feet altitude with the rain coming in all over the tent. At least once it was dropped and went rolling down the Khud [ravine]."[58]

However, despite the common assumption that under colonialism the office was what Sarkar calls a "man-only domain," use of the typewriter, like the sewing machine, raised significant questions about race and gender. In his South African law practice Gandhi employed two European female secretaries, though subsequently, on his return to India in 1915, he decided that it was unnatural for Indian women, life's nurturers and carers, to earn their living working outside the home as typists and receptionists.[59] Gandhi may have been representative of a wider prejudice, but it is clear from contemporary newspaper advertisements that a number at least of European women in India took up typing, whether to earn a living or help augment it. In the "situations wanted" columns of the *Bombay Chronicle* in July 1920 an advertisement appeared on behalf of a "well-connected English lady," with the "highest official references," who sought a position as "companion-secretary" or "companion-governess" and listed among her accomplishments "piano, singing, French, German, [and] typewriting."[60] The Government of India employed European and Eurasian women typists in the home department in Delhi up to and during World War II. Perhaps this was because they were required to have a good knowledge of English (at a time when levels of literacy among Indian women remained low), or, more likely, because they were trusted for racial reasons to handle confidential correspondence and work alongside officials who were themselves Europeans. But Europeans were not the only women to become typists in India.

In the United States in the late nineteenth and early twentieth centuries the typewriter fed the phenomenal growth of the modern office and women's employment. In 1870 women in the United States constituted a mere 4 percent of the 154 typists and stenographers; by 1900 this figure had risen to 77 percent of the total, and by 1930 to 96 percent (775,100 out of 811,200). There was a comparable surge in office work in Britain, where 146,000 women were working as clerks by 1911. There, too, the "professional opportunities opened to women through the medium of the typewriter played an important part in the movement for social equality between the sexes."[61] Did the typewriter have a comparable effect on women's employment and social position in India? Surely not. Unlike in the United States or Britain, the majority of typists in India remained men. But, since women, by virtue of their more nimble fingers and lower wage costs than men, had come to be seen as essential to office life in the West, it was perhaps inevitable that women should be expected to fulfill a similar role elsewhere. By the 1920s photographs of offices in India show European and Eurasian women wearing long white dresses, taking shorthand, answering phones, operating switchboards, or with hands poised over typewriters.

The number of women office workers was certainly small by Western standards. The 1931 census for Bombay recorded 105 women working in the city as "public scribes and stenographers" alongside 182 men, and entered a further 2,408 females employed as clerks, cashiers, and accountants alongside 50,016 males. In Calcutta at the same date there were 182 women stenographers (to 294 men) and in Madras 43 (to 477 men), though this latter figure represented a significant increase from a solitary female stenographer twenty years earlier.[62] If census categories can be relied on, across India as a whole it would seem that there were only a few hundred women typists by the 1930s, a modest beginning to the now widespread employment of women in Indian offices and call centers. This, though, was a situation in which, once again, gender was complicated by race.

Many of the typing and secretarial schools set up in Bombay, Calcutta, and Madras catered specifically to women. Vacancies appeared in the press for women typists, usually in private firms rather than government offices and often without reference to race. Many middle-class Indian women taught themselves to type in order to participate in the running of political and social welfare organizations; others did so to assist their husbands or for their own correspondence. But an important role in the training of women as typists and secretaries was played by the Indian branches of the Young Women's Christian Association (YWCA) and, for men, by the Young Men's Christian Association (YMCA). Like many other technological changes affecting India, this drew on American precedents. In 1881 the central branch of the YWCA in New York began teaching typing as a suitable occupation for young, single women. Twenty years later the Calcutta YWCA followed suit, offering typing lessons for women and these rapidly became popular: by 1905, 72 women were enrolled in its classes. Three years later the Calcutta YWCA moved to larger premises, which gave more room for typing and shorthand classes, "both of which the Association made a specialty." By 1920 the YWCA's "Commercial School" had 152 students and thirteen typewriters. Many of its "business girls" took up secretarial posts in Calcutta, including government offices: they were in particular demand during World War II, as Calcutta became a center of Allied military activity.[63]

The initiative of the YWCA in Calcutta was replicated, on a smaller scale, by secretarial classes in YWCAs elsewhere in India and by typing, shorthand, and bookkeeping classes for men at YMCAs. By the 1900s many European, Eurasian, and Christian schools across India had added typing and stenography to their curriculum, especially but not exclusively for girls, and were finding office jobs for their pupils. But the Calcutta YWCA retained a particular significance—its location allowed it to supply women office workers for one of India's leading commercial and industrial centers. It was also patronized by the wives of the viceroys and the lieutenant governors of Bengal (until the imperial capital moved to Delhi in 1912), who donated typewriters for secretarial training.

The Calcutta YWCA was clear about its feminist agenda. Typing and secretarial work was a means of giving women a decent occupation: it helped make them self-sufficient and gave them "the gift of power in their lives." The YWCA also targeted a specific kind of female constituency—Europeans from "home" (Britain), "domiciled" (India-born and India-educated) Britons, Eurasians, and Indian Christians. As long as colonialism lasted, this gave the secretarial enterprise a racial configuration. As with sewing machines, there was a racial assumption that white and mixed-race women were more amenable to modern technology than their Indian counterparts. They were assumed to be more literate (in English), better educated, and have the skills and disposition needed to operate a machine. Unlike most Indian women, they were free to work away from home in the socially and sexually problematic space of the modern office. The recruitment of Eurasian typists parallels their employment as nurses and midwives and shared the same rationale—a willingness to work outside the home, in forms of employment high-status Hindu women shunned.

There was, besides, a longstanding colonial concern to find suitable employment for Eurasians and domiciled Europeans who had, by dint of their racial identity, a close (if ambiguous) relationship with the imperial order. Largely debarred from the army and the upper echelons of the civil service, Eurasian men found alternative employment in a range of technology-related occupations—on the railroads as train drivers, guards, station masters, and clerks, and in the post, telegraph, and irrigation departments. However, by the early twentieth century Eurasians and resident Europeans faced increasing competition from Indians in many of these fields of employment, and the pressure to Indianize government services was mounting. At exactly the time that the YWCA and YMCA took up secretarial work as suitable employment for Eurasian and domiciled European women, a number of reports were demonstrating, especially for Calcutta, how their socioeconomic position was slipping. Some commentators blamed Indian competition; others emphasized what they saw as moral and physical laxity, a lack of drive and determination among Eurasians. Encouraging both sexes to become typists or stenographers was one way of sustaining the role of Eurasians as technological "go-betweens" while rescuing them from a life of indolence and immorality.[64]

In actuality, though, the outcome was somewhat different. Male Eurasians found themselves by the 1930s increasingly squeezed out of office employment. By contrast, in many private firms, especially those owned and managed by Europeans like Binny's in Madras, "the sphere of the typewriter was reserved exclusively for [female] Anglo-Indians."[65] New urban employment opportunities—as typists and telephone operators—helped create the image of the stylish, independent, and technologically savvy "modern girl" in India as elsewhere around the world. Expressive of this new appetite for modernity and glamour, an early Hindi-language film produced in Bombay bore the title *Typist Girl*.[66] But allocating a racially defined category of women to office work created its own, not always favorable, stereotype of the Eurasian woman. In Guru Dutt's 1950s movie *Mr and Mrs 55*, itself a cinematic paean to Bombay's technological modernity—the plot revolves around telephones, cameras, automobiles, taxis, and airplanes—there is a scene in which Johnny, a Eurasian photographer, pecks out the words "I LOVE YOU" on the typewriter of Julie, a Eurasian typist. They sing and dance around the office until other workers return and disturb them. The film ends morally, with Johnny and Julie going off to get married, though arguing whether it should be in a church or a registry office. But less sympathetic representations abound in which the Anglo-Indian woman typist is presumed to be morally loose and sexually available: Laurie Coutino in Rohinton Mistry's novel *Such a Long Journey* is one example of this. More negatively still, Violet Dixon, a character in a Mulk Raj Anand short story, is described as a "plain snub-nosed Anglo-Indian typist" with a "sing-song voice" and the "snobbish" air of the "memsahib she affected to be." As if to emphasize her half-alien identity, she boasts of having a married sister in Britain whom she wants to leave India to join.[67]

××××××××××××××

Like gender, ideas of race in British India acquired added authority by association with technology. "Higher" races were branded more adept at using modern technologies than "lower" ones. Parsi men and Eurasian women exemplified the way in which specific sets of people came to be identified with, even defined by, their relationship with modern machines. In the sociology of colonial collectivities, proximity to modern technology and evidence of technological aptitude served as a means of refashioning or reimagining entire communities, ascribing to them a privileged relationship to bicycles, typewriters, and sewing machines. Conversely, colonialism denied the eligibility of certain social groups to be modern by their apparent inability to operate and identify with those machines that were passing into everyday use. Even within the many gendered contexts in which everyday technology became colonially entangled, race could remain central, as in the case of Eurasian typists, to consideration of which operatives were most appropriate for particular kinds of machine-related work. In part this privileging of race and gender was illustrative of the peculiarities of India's colonial situation, and the particular preoccupation of India's rulers with the identification of highly differentiated social categories. But, as the social availability or denial of such modern goods as bicycles, sewing machines, and typewriters also suggests, this close identification of machines with gender and ethnicity was not a colonial invention alone but drew, too, upon notions of gender and community long established in Indian society.

CHAPTER FOUR

Swadeshi Machines

Until August 1947 India was an integral part of the British Empire. The way in which everyday technologies were imported into India from Western manufacturers, through the agency of foreign firms and, initially at least, with European customers in mind, would seem to reinforce the idea that these were essentially alien goods and constituted an expression of Britain's domination of the Indian economy. It might further be argued that the relative ease with which foreign goods entered, and largely monopolized, the Indian market, demonstrated the one-sided, exploitative nature of the imperial economy and the marginalization of Indian enterprise. The fact that it took until the 1950s for Indian-made bicycles, sewing machines, and typewriters to capture a leading share of the domestic market appears to support the nationalist assertion that the British ran India in their own selfish interests and stifled Indian competition—until India won political freedom and seized control of its economic destiny.

It is not the intention here to contest this broad-brush portrayal of colonial domination and exploitation. But it is necessary, from the perspective of everyday technology, to add some qualification to this argument. For a start, as previous chapters have shown, a significant proportion of the technological goods *due to free trade* entering India in this period came not from Britain but from American and European manufacturers, albeit through conduits of trade and governance the British had established. It is necessary, too, to consider ways in which Indian enterprise, despite the many obstacles in its path, was able to gain a degree of control over the production, and more especially the distribution and sale, of imported commodities, thereby gaining a commercial foothold that could be extended once a more favorable economic environment had been created after World War II. And, since our concerns are not primarily economic but social and cultural, it needs to be recognized that there were many ways in which Indians established a sense of ownership over and identity with technological goods that were, at least originally, of foreign provenance. To understand this emerging sense of ownership we need to turn first to the concept of *swadeshi* and its role in the dissemination and appropriation of everyday technology.

The Idea of *Swadeshi*

The conventional view of *swadeshi* is easily stated. In the latter years of the nineteenth century Indian intellectuals, led by Dadabhai Naoroji and Romesh Chandra Dutt, took the view that their country was being drained of its wealth and reduced to chronic poverty by British rule. This was attributed to heavy administrative costs charged to India, military expenditure that only partly related to Indian defense, high salaries and pensions for European civil servants, and inequitable trade policies. Combined with the burden of land revenue demands and the seemingly precipitous decline of once flourishing industries (like the cotton textiles for which India had formerly been renowned), these imperial charges and exactions had, it was argued, brought India to a pitiful state of penury. Indebtedness, landlessness, famine, and pestilence reduced the well-being of the population still further. For critics like Naoroji this was as much a "moral loss" as evidence of the "material exhaustion" of India.[1] Apart from reducing or annulling current costs and charges and scaling down revenue demands, one proposed solution to the continuing "drain" of India's wealth was to raise import duties to give tariff protection to infant industries (like the textile mills of Bombay). However, Britain's strict (and self-serving) adherence to the economic principles of laissez-faire and the political influence of British manufacturers, with a few exceptions (like sugar production in the 1930s), precluded this from happening. India remained exposed to the importation of exogenous goods, though, as with bicycles and typewriters, this might mean American, German, Japanese, and not just British, commodities.

Even before the term *swadeshi* gained wide circulation, the idea of attacking British imperialism through its economic interests and giving encouragement to the creation or revival of Indian industries had become a potent part of the nationalist agenda. Drawing inspiration in part from colonial exhibitions, Indian patriots began to organize counterdisplays of Indian crafts and artifacts as a way of instilling national pride and demonstrating how India might recover its former prosperity.[2] While the emphasis in these exhibitions tended to be on restoring what colonialism had destroyed, the desirability of looking forward to India becoming a modern industrial nation was also vigorously promoted. As the nationalist movement, spearheaded from 1885 by the Indian National Congress, grew more assertive, especially following the partition of Bengal in 1905, more radical responses were proposed. One was the boycott of British goods (especially the textiles so central to the imperial economy); another was the creation of new Indian industries, funded by Indian capital, to compete with foreign firms and to manufacture Indian goods for Indian consumers. This economic strategy sought to undermine British trade while returning to India "her rightful, her ancient place, among the industrial nations of the earth."[3]

At its most radical, as espoused by Aurobindo Ghosh in Bengal, the *swadeshi* movement was directed "not merely against foreign goods, but against foreign habits, foreign dress and manners, foreign education," and so sought to bring the Indian people "back to their own civilization."[4] The movement aspired not only to end Indian dependence on foreign goods but also to promote "national regeneration" by expunging an alien way of life.[5] Some, like the writer Ananda K. Coomaraswamy, went further, seeing in the movement an opportunity to attack the "vulgarization of modern India." The neglect of India's artistic refinement and vibrant craft tradition had, in his view (not unlike that of George Birdwood a few decades earlier), been accentuated by the adoption of foreign goods and tastes, from clothes and carpets to gramophones and harmoniums. Modern machines had not enriched Indian life, only coarsened and impoverished it.[6]

The origins and progress of the *swadeshi* movement in its political heartland, Bengal, have been detailed by Sumit Sarkar, whose discussion shows the broad economic parameters of the movement and the resurgent cultural nationalism it helped provoke. But he devotes particular attention to the failure of the movement which, by his account, foundered on the opposition of British commercial and industrial interests (backed by the colonial state) and on the lack of entrepreneurial skill, technical expertise, and capital resources among the Bengali *bhadralok*. It is seen to have suffered, too, from the inability of a high-caste Hindu leadership to engage effectively with the lower classes, including the Muslim peasantry of east Bengal. The collapse of the *swadeshi* movement was compounded, Sarkar argues, by the drift into terrorism in 1908 that further detached middle-class revolutionaries from the masses and met with severe state repression. The movement thus appears relatively short lived and largely unsuccessful, at least as far as its grand ambitions were concerned. Sarkar's own sense of "anticlimax" at the movement's limited achievements parallels the view of the Indian Industrial Commission in 1918 that attributed the failure of *swadeshi* "almost entirely . . . to lack of business aptitude and commercial and industrial experience in classes which had had no opportunity of acquiring them."[7] The movement did, though, in Sarkar's opinion, serve as a precursor for the more successful mass movement led just over a decade later by Gandhi, in which many aspects of the earlier *swadeshi* program, including boycotts, noncooperation, the making of homespun cloth, and the revival of cottage industries, were adopted as part of a nationwide campaign for Indian freedom. But, for all the inspiration and impetus it gave to subsequent nationalist agitation, by 1911, when the British annulled the partition of Bengal and transferred the capital of British India to Delhi, the movement was, in Sarkar's view, largely over.

The Bengali intelligentsia certainly provided much of the initial thrust behind the *swadeshi* movement, but it was taken up in other parts of India as well—in north India and Punjab, in Maharashtra in the west, and in the Tamil- and Telugu-speaking south. Moreover, although many of the advocates of *swadeshi* urged the creation of major industries to drive out foreign competition, in actuality it was often small-scale enterprises, making modest items of everyday consumption, that were best able to thrive. A *Directory of Indian Goods and Industries*, published on behalf of the Congress in 1909, listed a large and diverse range of enterprises across India that called themselves *swadeshi* and sold such items as soap, matches, and ironmongery. One of the most successful of all *swadeshi* enterprises was the production and sale of Ayurvedic medicines, a primary feature of Hindu cultural revivalism but also a significant avenue for emerging Indian commerce and consumerism. Even if, despite the boycotts, many foreign goods retained their markets and their prestige, entries in the directory suggest (as does a great deal of contemporary advertising) the growing importance attached to the *swadeshi* label. It had powerful patriotic connotations, an association with ideas of cultural and physical purity (as in the manufacture of soap free from animal fat), and made an appeal to indigenous tastes and "moral consumption." It catered to an Indian (especially high-caste Hindu) sense of appropriate modes of dress and bodily practices and fueled the reaction against what were seen as decadent and unpatriotic Western tastes. Machines could have *swadeshi* credentials too. The movement encouraged Indian interest in developing small-scale technology (such as the indigenous manufacture of rice-husking machines) and in seeking patents for locally devised and innovative machines.[8]

Recent scholarship has begun to reveal the *swadeshi* movement's contribution to revolutionary changes in Indian consumer tastes and demands.[9] It is clear, from even a casual reading of contemporary newspapers and trade directories, that *swadeshi* idealism and the market for India-made goods did not evaporate after the collapse of the political movement around 1908. It continued to survive, for cultural as much as economic reasons, and there was a strong revival of *swadeshi* enterprise in the late 1930s. Nor was *swadeshi* an exclusively nationalist ploy. Once the idea had gained popularity, even American cigarette firms and British soap makers began to establish their own factories in India, and, like Dunlop tires, to promote their products as also being "made in India, by Indians, for Indians."[10] While many of the pioneers of the movement had hoped, ambitiously, to create full-fledged industries to rival those of the industrial West, and thought of *swadeshi* as essentially a revolution in manufacturing, others recognized that Indians first needed to take control of marketing and distribution. The Irish-born patriot Margaret Noble, known as Sister Nivedita, addressed this issue early in the movement's history. Foreseeing serious obstacles to the creation of *swadeshi* industries, she urged that "the channels of distribution and the small shops—which are the *real* distributing centres in every city—have been so long in the hands of the foreign trade that they require to be captured now." These were the places where the schoolboy bought his paper and ink, the housewife her daily utensils. Only by capturing the "four-anna shop" and "the storeroom of the poor" would *swadeshi* ultimately "penetrate . . . into the remotest . . . villages and huts."[11]

Cycling toward *Swadeshi*

As chapter 2 showed, about 2.5 million bicycles were imported into India during the colonial period, most of them originating in Britain. To familiar brand names like Raleigh and Hercules were added cheaper, if less fashionable, machines manufactured in places like Coventry and Glasgow specifically for the export market and that, by virtue of their lower price, had a significant role in popularizing cycle use in India. As early as 1917 a British official, Frederick Nicholson, suggested to the Indian Industrial Commission that bicycles were one of several modern goods that could easily be produced in India. Little capital was required, the component parts could be manufactured in small workshops (as they still often were in Britain), and, without the addition of shipping costs, cheap Indian bicycles could both be popular and stimulate further demand.[12] In fact, because of the strength of foreign competition, the absence of protective tariffs, and the slow progress of precision engineering in India, it was more than twenty years before Nicholson's expectations even began to be fulfilled.

FIGURE 4.1. A cycle repairman at work in Pondicherry (Puducheri) in 2008. The street remains an essential site for the maintenance everyday machines in India. Author's photo.

However, even if India lacked the technical capacity and market opportunities to make and sell substantial numbers of whole bicycles before the 1950s, it was still able to participate actively in processes of assembly, sale, and distribution. This was done partly through the assembling and repair of bicycles using imported parts. In 1907 a prospectus announced the formation of a "National Cycle and Motor Company." It aimed to start a "new industry" in India, though in practice, for the present, its ambitions were limited to the more modest task of importing and repairing foreign bicycles.[13] By 1911 Calcutta had two workshops, employing 44 workers, for assembling bicycles. Twenty years later more than a thousand people in Calcutta and over 2,500 in Madras were employed in the repair of automobiles and bicycles. Lahore reputedly had its own "bicycle makers."[14]

It is impossible to calculate exactly how many machines were put together in this way, but during the interwar years the value of imported cycle parts was often more than half the value of whole machines; in some years (as in 1929–30) it even exceeded it. This suggests that in addition to a constant need for replacement parts, there was a lively trade in frames, handlebars, and other components that could be used to complete or assemble bicycles in India. Like the automobile industry, equally unable to establish itself in India until after independence, importing and assembling component cycle parts provided a base for what, after 1947, became a major industry. Imported cycle frames, chains, and saddles, like the parent machines, came mostly from Britain, but a significant number also came from Germany, Japan, and the Netherlands, giving local retailers the opportunity to market machines that were not solely imperial products. (One of the debates within the *swadeshi* movement was whether in boycotting British goods, Indians should still import manufactures from other countries until such time as India could make its own; the bicycle trade suggested the validity of this argument.) Some components, such as lamps, bells, stands, and carriers, began to be manufactured, as virtual cottage industries, in and around Ludhiana in Punjab by the late 1930s.[15]

While discussion of *swadeshi* has mainly focused on Bengal, the role of Punjab, a thousand miles to the northwest, was also highly significant. The province was not only a leading market for imported sewing machines and bicycles but also a source of technological innovation and enterprise. The close connection with the army was one factor in this (a large part of the Indian Army was recruited from Punjab), especially in providing experience of machinery, such as driving and repairing military trucks. Punjabis, and Sikhs in particular, became extensively involved in operating or renovating all things mechanical. By the 1920s, many of the taxi drivers in Calcutta were Sikhs, as were many truck drivers across northern India.[16] Bishandas Basil, the engineer who developed the first Indian-made sewing machine (the "Usha"), worked in Calcutta but was originally from Ludhiana in Punjab.

Bicycles became available, not just through the large numbers that were imported whole into India but also through those that were assembled or manufactured there. However, in a situation in which the main site of manufacture remained outside India,

FIGURE 4.2. An undated photograph, probably late 1930s, showing cycle shops at Gwalior in Central India. Reproduced from the Raleigh Archives (DD/RN 6/19/4) by kind permission of the Nottinghamshire Archives, Nottingham, UK.

distribution and sale were almost as crucial in giving Indians a share in the market and a sense of ownership over new technologies. In cities like Calcutta, Bombay, Madras, and Lahore, European department stores and agencies sold cars, motorcycles, bicycles, typewriters, gramophones, and other imported goods, initially for a European clientele. But, by the 1910s and 1920s, Indian retailers were also widely involved in these trades and in some instances drove European competitors out of the market. Indian cycle dealers appeared on the streets of almost every town and city, selling their wares in open-fronted shops or on the pavement. Many such firms also sold cycle parts and accessories, undertook repairs, or combined the sale of bicycles with that of motorbikes and automobiles. The Indian bazaar was critical to this burgeoning activity.[17] A site for establishing familiarity with such novel objects as sewing machines, the bazaar also became a place where new, but also secondhand and reconditioned, machines—bicycles, sewing machines, typewriters, gramophones— could be bought or hired. With its blend of novelty and excitement, of desire and display, the bazaar represented a site of cultural exchange as well as commercial transaction.

By the 1920s Calcutta had nearly thirty cycle dealers and repairers, many of them located in the intermediate commercial zone between the European and Indian sections of the city, along Bentinck and Dhurrumtolla Streets. Madras similarly had forty cycle importers, dealers, and repairers, most of them Indian, located in the commercial area of George Town or around the smart shopping locality of Mount Road. Delhi in 1935 had sixteen cycle dealers, mainly around Chandni Chowk in the old city, with others located around Connaught Place in the new imperial capital.[18] Just as Parsis were involved in the retail sale of sewing machines, so were they with bicycles. Firms, such as Rustomji and Co., were active in the cycle trade with branches across western and northern India. The first cycle dealership in India was that of the Bombay Cycle Agency run by a Parsi, K. D. Wadia, in 1885. Later renamed the Bombay Cycle and Motor Agency, it imported, serviced, and repaired automobiles. It was eventually taken over by a Gujarati industrialist, Walchand Hirachand, as part of his attempt to create an Indian automobile industry, a pattern of industrial progression that had close parallels in Britain and the United States.[19]

Well before independence, Indians also became leading entrepreneurs in the cycle trade—as the case of Sudhir Kumar Sen demonstrates. Born in 1888, Sen came from a well-educated and progressive *bhadralok* family in Calcutta: his sister Yamini was one of the first Indians to enter the Women's Medical Service when it was formed in 1914. His interest in bicycles began as a student at Presidency College, where he won an intercollege cycle race in 1905.[20] Sen was influenced by the patriotic idealism of Prafulla Chandra Ray, a pioneering scientist and the founder of a thriving chemical and pharmaceutical works established in Calcutta in the early 1890s, who appealed to Bengalis to set up *swadeshi* enterprises. Ray became an ardent supporter of Gandhi, denouncing what he saw as the swamping of India by foreign "luxuries" such as automobiles, bicycles, and harmoniums.[21]

Eschewing Ray's asceticism and hostility to foreign machines, Sen interpreted *swadeshi* in his own way, seeing the bicycle not as an unpatriotic indulgence but as a means to Indian well-being and self-sufficiency. For him, promoting the use of bicycles, even by marketing foreign machines, was a necessary step toward eventually making them in, and for, India. He began in 1910 importing and selling bicycles in Calcutta through a firm called Sen and Pandit, with offices on Bentinck Street. Following a long legal dispute with the British firm of Oakes and Co., in 1919 Sen won the right to sell cheap British-made "Warrior" bicycles, costing only £2 (about Rs 40 at the time).[22] Sen and Pandit expanded to become the largest cycle importers in India, holding the agency for BSA, Hercules, and Raleigh machines. By the late 1930s nearly half of all bicycles imported into India passed through Sen and Pandit. Eager for non-British suppliers, Sen traded in German saddles and imported Japanese bicycles but the latter never sold well.

Not content with his dominant position in the import trade, Sen continued to pursue the idea of a truly *swadeshi* bicycle and sought to popularize cycling in India, especially for the masses. Through small workshops in Bombay, Calcutta, and Punjab, an indigenous bicycle-making industry began to develop in India in the 1930s, but it was held back by the war and the absence of an Indian capacity to manufacture high-quality steel for chains, ball bearings, and freewheels. Only around 44,000 machines were made in India in 1945, despite an estimated demand for half a million cycles a year. Faced, after the war, with an unprecedented influx of foreign machines, Hind Cycles in Bombay, which by the late 1940s was producing 150,000 machines a year, stressed its *swadeshi* credentials, appealing to customers to buy bicycles that were "built in India, built for India, built by Indians."[23] High import tariffs and a state policy of encouraging import substitution in the late 1940s and early 1950s began to stifle the booming trade British companies, led by Raleigh, had hitherto enjoyed.[24]

Even so, despite the passing of colonial rule, many of the technological constraints on Indian production remained, and local firms found it necessary to seek foreign collaborators, including British manufacturers. In 1950 Sen signed an agreement with Raleigh by which the British firm supplied technical expertise to a joint Sen-Raleigh company and helped build a modern bicycle factory in rural West Bengal. The aim was to manufacture

100,000 bicycles a year—a fifth of the National Planning Committee's annual target but only a tenth of the one million machines Raleigh was currently producing in Nottingham for the UK and overseas markets.[25] But Sen (who died in 1959) was not the only contender. With assistance from Tube Investments of Birmingham, a rival cycle factory was established near Madras in 1951 with a similar target of 100,000 bicycles a year. By the mid-1950s Hind Cycles and Atlas had also become leading manufacturers giving India a combined capacity of 400,000 machines a year. The prejudice against Indian-made machines was said to have been overcome, and every additional cycle sold helped "advertise them on the road."[26] The shift to indigenous production represented significant industrial progress, but it was not unqualified. The industry relied heavily on foreign collaboration and was geared to making locally an already familiar global commodity rather than fostering innovative research and new forms of technological development.

Even so, the rise of the Indian cycle industry, from colonial beginnings in the 1890s to national control in the 1950s, shows the importance of Indians gaining a hold over an area of everyday technology that was initially wholly dependent on imported goods and European agency houses, but which became increasingly indigenized. It demonstrates, too, how, despite the foreign origins of the machine and colonial constraints on trade and manufacture, Indians were able to develop their own patriotic rhetoric around the bicycle, and came to see it as an appropriate vehicle for their *swadeshi* ambitions. This paralleled the growing popularity of bicycles in India. Well before independence the bicycle had moved beyond middle-class use. It was a mark of the bicycle's growing availability that while more expensive Raleigh machines were dubbed the "babu" (gentleman's) bicycle, the cheaper Hercules became the "mazdoor" (workman's) bicycle.[27]

Among the few modern goods low-caste villagers had begun to acquire by the 1960s, bicycles figure prominently in stories told, in fiction or as life narratives, about the opportunities and conflicts of everyday life. One of Mulk Raj Anand's short stories relates how a barber, tired of being at the beck and call of the high-caste men of his village and being poorly remunerated for his work, buys a secondhand Japanese bicycle for Rs 5. Once he has mastered the art of riding it, the bicycle enables him to travel to the nearby town in search of better paid work. In time he is able to sever all links with the village and set up his own barber-shop.[28] But there were social, as well as economic and technological, obstacles to the bicycle's spread. Acquisition or use of a bicycle by low-caste villagers could provoke bitter opposition from the higher castes. In her account of Tanjore in the early 1950s, Kathleen Gough describes how a villager from the low-ranking weaver caste, a "rebel who rode a bicycle," collided with a high-caste boy and was obliged to pay a fine of Rs 100 as compensation for his injuries.[29] But, even so, as a machine (in this subaltern context, less one of brazen modernity than of quiet utility), it was hard to prevent the growing use of bicycles among low-caste and untouchable communities. James Freeman, in his account of a village near Bubhaneshwar in Orissa in the 1960s and 1970s, noted how men of the washerman caste used bicycles to carry laundry to and from the city and how a man from the fisherman caste had successfully set up a cycle-repair shop. Even members of untouchable castes began to use bicycles, though it remained an important event when one of them first acquired a cycle.[30] *duality*

The *swadeshi* movement thus had multiple meanings. For some it essentially signified India's capacity to become a self-sufficient industrial nation; for others, like Gandhi and P. C. Ray, it was more about rejecting foreign machines and unpatriotic lifestyles and returning to traditional goods and tastes. However, for a group of enterprising Indians, like Sudhir Kumar Sen, *swadeshi* encompassed an ultimate ambition to manufacture modern industrial goods in India as well as a more immediate aim of capturing the assembly and distribution of imported goods and of putting them to the service of the Indian population at large.

In Search of the *Swadeshi* Typewriter

Like other increasingly everyday machines, typewriters were widely advertised by their brand names in India's English-language press. They appeared, alongside bicycles, sewing machines, cameras, gramophones, and other household goods, in newspaper advertisements and in "wanted" and "for sale" columns. Rebuilt and secondhand machines also retailed at substantially lower prices. New typewriters were mostly sold through the European or American agents of Remington and other leading firms, but Indian companies held a large share of the retail, secondhand, and repair market. As with sewing machines and bicycles, an ancillary industry grew up supplying typewriter parts, ribbons, and stationery. Some of these enterprises were at pains to stress their patriotic credentials. "True sons of India," ran the appeal from one Agra firm, selling pen nibs, leather goods, and buttons, "Your desire for the uplift of this sacred land can be achieved only by using nothing but Swadeshi articles."[31]

Like bicycles, a large proportion of typewriters sold in India in the 1910s and 1920s were sold by Indians—as sales agents for Remington and other foreign firms, or through the many Indian distributors that existed by that time. The Madras *Almanack* for 1912 listed twenty-one dealers and agents in that city alone, most of them with Indian proprietors and located in, or close to, the commercial heart of George Town. Bombay by the same date had close to thirty European and Indian dealers, selling British, American, German, and other foreign machines.[32] But, as with bicycles, Indian entrepreneurs and idealists held the *swadeshi* belief that India should do more than simply sell foreign typewriters. It should ultimately produce its own typewriters—despite the complex machinery and precision engineering required. India, after all, was not just an outlet for conventional typewriters with a Roman script keyboard but also potentially a market for those that used Devanagari and other Indian scripts. It is unclear when the first non-Western keyboards were produced—or by whom—though there is some tantalizing evidence. In 1910 the Government of India's home department asked for a supply of Hammond typewriters with "interchangeable vernacular alphabets" for use in Calcutta. An American history of typewriters published in 1923 makes reference (without giving its source) to "The Hindu [sic] vernacular machines, especially the Marathi,"

FIGURE 4.3. A Remington typewriter salesman in south India, c. 1910, a picture taken by the celebrated photographer Deen Dayal. This image is reproduced from *Princely India: Photographs by Raja Lal Deen Dayal, 1884–1910*, ed. Clark Worswick (London: Hamish Hamilton, 1980), 118.

which were "having a considerable sale today among the native princes and potentates of British India."[33]

India's princely states and landed elites encouraged, perhaps pioneered, non-Roman scripts for typewriters just as they provided employment for shorthand clerks and secretaries using Indian languages. In January 1920, for instance, the maharaja of Alwar in Rajputana sought a private secretary with a knowledge of English, Hindi, typing, and shorthand. The following month, the manager of a *zamindari* estate in the United Provinces offered for sale (at the hefty price of Rs 550, just over £27 sterling) a barely used "Remington Dev-Nagri typewriter" in good working order.[34] In a significant move toward vernacularization, even before the end of World War I, Remington was producing typewriters adapted for the use of the Urdu, Marathi, Gujarati, and Bengali scripts.[35] By the 1930s Hindi typists had become a regular part of the government establishment in northern India, so by then use of Devanagari keyboards must have been fairly widespread.

In India, as in Europe, the United States, or Mexico, the typewriter, along with the camera, gramophone, and radio, became an exemplar of technological modernity—a marker of modern state power but also of modern artistry and self-expression. It was also, as Friedrich Kittler and Rubén Gallo have shown in other contexts,[36] part of an extended family of modern machines that, in their technological, commercial, and social interconnectedness, spoke to one other. Today it is possible to detect, at least in retrospect, a singular affection for the typewriter and its disappearing band of users. However, in colonial India this passion for modern technology in general, and typewriters in particular, was more muted than in many other parts of the world. A "craze" for the modern was constrained not only by endemic poverty, by foreign monopolies, and by patriotic concerns but also by a critique of the machine as soulless, impersonal, and foreign. Thus Gandhi, who had employed typists in his law firm in South Africa before 1910, and was a sufficiently accomplished typist to teach others, opposed the typewriter by the time he returned to India in 1915. He later expressed his "dislike of typewriters," claiming that the machine was "a cover for indifference and laziness." It had all but destroyed the "magnificent art of calligraphy" and threatened to replace the handwritten manuscript through which the writer poured his "very soul" into his work.[37] It is possible to find Indian writers, from R. K. Narayan to Ruskin Bond,[38] for whom the acquisition of a typewriter marked a crucial stage in the development of their literary career, but many writers, working especially in the vernaculars, seem to have preferred writing by hand, leaving the typewriter to the humdrum clerk.

Even so, vernacular typewriters might hold a special place in the nationalist imaginary. Typewriters could promote a national language as well as serve a colonial one. Partly by virtue of its distinctive language requirements, the domestic market provided a stimulus for indigenization, especially when in 1950 the Indian constitution recognized Hindi in the Devanagari script as the official language, allowing English only fifteen years' respite before it was replaced by Hindi. Although English was later granted a reprieve and has continued in official use, the adoption of Hindi as the national language made it politically imperative to have a reliable supply of Hindi typewriters for use in government offices. In the constituent states of the Indian Union, there was a comparable need for typewriters using Tamil, Marathi, Gujarati, and other Indian scripts, and, until production recently ceased, government offices were the leading market for these vernacular machines.

The principal firm to take up the challenge of manufacturing an Indian typewriter was that of Godrej and Boyce in Bombay. We will return to the history of this company in the next chapter in discussing the relationship between technology and health. It is sufficient for the moment to note that this Bombay firm was founded in 1897 and advanced from an early interest in surgical instruments and soap (a leading *swadeshi* commodity) to making locks, fireproof safes, steel cupboards, and office furniture, goods that spanned the spectrum of domestic, commercial, and industrial use and put the firm at the forefront of *swadeshi* consumerism. As further evidence of its nationalist credentials, the firm made substantial contributions to the Tilak Swarajya Fund set up by Gandhi in 1920. In 1939, on the eve of World War II, Godrej intended to move into bicycle production (a further illustration of how interconnected the manufacturing and sale of everyday goods could be), but when the ship carrying the machinery for its cycle-making plant was sunk by a German U-boat they abandoned the idea, contenting themselves instead with supplying chains and tubing for the *swadeshi* firm of Hind Cycles.

With the demise of British India, and the formation of a national government under Jawaharlal Nehru, a keen exponent of economic autarky, Godrej was an obvious choice for developing an India-made typewriter. As early as 1942 it had contemplated entering the typewriter market but lacked the expertise and resources to produce a *swadeshi* machine. The technical difficulties of producing even an apparently simple machine like a bicycle, let alone a sewing machine or typewriter, presented a continuing obstacle to indigenization. In keeping with the imperial need for grand "state works," under colonialism, India's engineering expertise mainly lay in large-scale projects, such as bridge building and railroad construction. This had not been matched by the precision engineering required to make small but complex items of machinery. While it was possible to replicate many of the basic components of foreign machines, India into the 1950s remained seriously deficient in machine tools, a significant hindrance to the production of goods like bicycles and sewing machines that relied on precision engineering for their effectiveness.[39] A sewing machine needed around 220 components able to perform 1,800 separate operations.[40] Making a typewriter with 1,800 constituent parts also called for high-grade steel, iron, aluminum, brass, rubber, and plastic, few of which were available for civilian use in wartime India or during the immediate postwar years. As Godrej's historian explained:

> In addition to the usual turning, drilling, tapping, riveting and punching jobs, typewriter manufacture involves the making of over one hundred and fifty varieties of screws with special threads, not to speak of a whole series of specialized jobs like die-casting, spring-coiling, rack-cutting, gang-milling and heat-

treating. [It also requires special milling machines for making] type-bar segments, carriage guides and mounting frames; die-casting machines for main frames, top plates and spool covers; automatic rack-cutting machines for carriage racks, tabulator stop and margin stop bars; multiple spindle drilling and tapping machines for key lever mounting, carriage and plates and other components which require precision drilling.[41]

With political support from Nehru and encouraged by guaranteed government purchases, Godrej began work on typewriters in the early 1950s. Its first commercial product, patriotically named the "All-Indian," appeared in 1955. Just as the prototype Indian sewing machine, the "Usha," took a recent German machine (the Pfaff) as its model, so the first Godrej machine was based on an American "Woodstock" typewriter. Subsequently, however, as India looked to the Soviet bloc for its technology, Godrej turned to an East German firm for components and expertise. This uneasy collaboration lasted until 1967. Only then, twenty years after independence, was Godrej able to produce machines competent enough to rival foreign makes. Even then, like the first Indian sewing machines, the appearance was "shoddy," with poor quality paintwork and finishing, and with component parts that were apt to break or wear out quickly. Godrej's first typewriters had a "hard touch," were clunky and tiring to use, and, as the reports of their sales representatives attest, met with frequent complaints from purchasers and users.[42] Only a combination of state commitment to indigenous production, tariff barriers against foreign imports, and growing technical skill enabled Godrej to increase its market share. Between 1955 and 1959 it sold only 9,000 typewriters, rising to 12,000 in 1959–1961. Between 1965 and 1970 production rose to 24,258 machines, and Godrej began to make typewriters with various different vernacular keyboards.[43]

But, for all the technical limitations of these *swadeshi* typewriters, they served an important patriotic purpose. A widely circulated photograph showed Nehru at the annual session of the Indian National Congress in December 1955 using a newly produced Godrej "All-Indian" typewriter. And so, the firm's official historian enthused, "The nation received the message that India was taking its place among the few highly industrialized countries of Europe and America capable [of] typewriter manufacture." "India," he added with pride, "was the first such in Asia."[44]

Modern Milling

Although the Indian countryside did not suddenly awake to the clamorous sound of steam plows and mechanical reapers as nineteenth-century British champions of "improvement" had hoped, there was a steady, in some ways remarkable, spread of agricultural machinery, especially machines used to process such agricultural commodities as sugar, rice, and oilseeds. Initial British attempts to introduce improved sugar-crushing machines apparently yielded little positive response. There was general skepticism among colonial officials as to whether technological change would ever penetrate the Indian countryside. The few machines that were sold were said to have been left to rust or abandoned for want of maintenance. "The prosperity of Indian agriculture," ran a 1911 report, "is closely connected with the efficiency of the blacksmith, and one reason why improved machinery cannot be introduced into the Punjab villages is the difficulty of getting it properly repaired."[45] According to W. H. Moreland in 1920, it was "common knowledge" that Indian workmen were "very apt to spoil modern tools and machines."[46]

However, the reality was somewhat different. By the 1880s thousands of relatively cheap, easy-to-use sugarcane crushers, equipped with metal rollers, were being sold to cultivators in Bihar, the North-Western Provinces, and Bombay. Indeed, they were soon said to "bid fair to drive the creaking wooden *ghani* [sugar mill] out of the field."[47] Among the most popular of these machines was the "Bihia mill" made by a British firm, Thomson and Mylne, in Bihar.[48] Light enough to be moved from field to field, the mill had parts that could be repaired by village blacksmiths, and it was sturdy enough to last ten harvest seasons. Indian entrepreneurs, like Subrav Chowhan in the Bombay Deccan, also took up the design and manufacture of sugar mills with two or three iron rollers. Although the cost of these machines (Rs 200 in the late 1880s) was too prohibitive for most cultivators to purchase outright, they could (like so many other increasingly everyday technological goods) be had for hire, for a rupee a day.[49] The value of the new machines lay in their greatly increased efficiency, which enabled a far larger quantity of juice to be extracted from the cane. Their use encouraged the spread of higher yielding varieties of cane and the growth of sugar as a commercial crop and mainstay of emergent agrarian capitalism. There were, of course, losers as well as gainers. Conventionally, in return for making the pots in which cane juice was stored, potters had the right to the crushed cane stalks and any residual juice squeezed out of them. With more efficient cane crushers there was little left to extract. By the 1920s sugar factories were appearing in many parts of the Indian countryside, to the undoubted benefit of their owners but to the questionable advantage of the peasants who produced the labor-intensive crop.[50] Despite the Depression, sugar, aided quite exceptionally by state protectionism, was one of the few agricultural commodities to flourish; both the importation and indigenous manufacture of cane-crushing equipment energetically kept pace with it.

In similar fashion the pressing of oil-bearing seeds to produce oil for cooking and lighting had once been the exclusive preserve of oil-pressing castes and the bullocks that turned wooden mills (or *kohlus*).[51] But they too faced competition—from imported kerosene and power-driven presses. As steam power, and subsequently oil and electric engines, became more widely disseminated, so mills to grind wheat flour and pulses also sprang up in cities and towns. But, to return to one of our principal examples, rice husking, the essential means by which raw paddy was converted into edible rice, and once one of the most widespread of all domestic tasks, was also being replaced by the modern rice mill.

The hopper and the rotating grindstone or perforated metal drum used for mechanical rice milling, like the attendant oil engine, were widely available in India by the 1920s from British and American companies, but they were also manufactured by Indian engineering firms in Bombay, Calcutta, and elsewhere. More feasible, given India's limited engineering base, than bicycles or typewriters, the invention and manufacturing of rice-husking machines became a significant element in *swadeshi* enterprise.[52] By the 1930s Madras had more than a dozen firms or their agents specializing in the sale of rice-milling equipment; some had their own foundries making rice and flour mills, oil presses, and machines for shelling groundnuts. The introduction of Indian machines, their design pragmatically pirated from American Engelberg hullers, significantly reduced purchase costs and facilitated the spread of rural rice milling. Although some distributors provided clients with written operating instructions, or sent a mechanic (a European or Anglo-Indian) to explain the working of the machine, in the main each operator learned how to become a "driver" (as the person in charge is still called in Tamil) by watching someone else do the job and lending a hand—helping to load buckets of grain into the hopper, checking the quality of the milled grain, cleaning and oiling the moving parts. Neither literacy nor formal training was required for this induction into modern technology. In stark contrast to earlier colonial complaints about Indians' technological ineptitude, by 1940 it could be stated as "a matter of common observation" that "the South Indian is quick to appreciate mechanical devices and that ability to handle and repair machinery is widespread."[53] To judge, too, from present-day observation in Tanjore, many of the operators came from precisely the kinds of artisan castes—blacksmiths and carpenters in particular—who formerly made and repaired such rural implements as sickles, plows, and hoes.

The great majority of south Indian mills were set up and owned by local entrepreneurs, members of "rich peasant" and trading communities for whom a move into rice milling and similar activities like cane crushing, groundnut decortication, and oilseed pressing formed part of a shift away from relatively unprofitable agriculture and petty commerce into areas of trade and manufacturing that promised new wealth and status. For instance, the Ellore Rice Mills in Kistna District, dating from 1900, were owned by Mothay Gangu Raju, who was also the proprietor of a jute and cotton mill. By 1914 the factory processed up to a thousand bags of paddy a day during the harvest season, using three steam engines to boil, hull, and polish the rice. In the nearby port of Cocanada (Kakinada) the firm of Innes and Co., which had begun as a European concern but from 1908 was run by K. Suryanarayanamurti Naidu, processed and exported rice, sugar, and other agricultural products. According to seasonal demand, the works employed between fifty and two hundred laborers. Another Naidu enterprise, the Coringa Company, fourteen miles from Cocanada, consisted of two rice mills, handling a thousand 166-pound bags of paddy a day and employing two hundred workers.[54]

With respect to rice mills and many other innovative technologies, naming practices were important in creating a sense of cultural ownership and indigenous proprietorship over once- or part-foreign objects. Loan words and vernacular renderings such as *saikal* (for cycle) or *taip-raitar* (for typewriter) were rapidly absorbed into Indian languages, helping to obscure an alien provenance. If some machines were known by their international trade names, like the ubiquitous "Singer," others were named to emphasize local associations. Thus "Usha" sewing machines were named after the daughter of the Punjabi engineer who pioneered their development. Bicycles clung rather longer than sewing machines to their Western names and the prestigious or adventurous associations they invoked. For a long time Sen-Raleigh's best-selling cycle was called "Robin Hood," and even today foreign names like "Hercules," "Atlas," and "Hero" remain common for bicycles. Some rice mills, too, simply bore place or family names. But the greater number, especially in south India, emphasized by their names the sanctity and purity of the life-sustaining food they processed. Naming practices identified factories with divine benevolence and the deities of which the owners were devotees—like the Sri Venkateswara mill in Cocanada or the Mahalakshmi Boiled Rice Mill in Bezwada—transforming them, if in name only, into industrial analogues of Hindu temples.

Ownership of a rice mill and involvement in agro-industries might bring prosperity to a new class of entrepreneurs, and part of this new-found wealth was channeled into ownership of other conspicuously modern goods, such as automobiles, or invested in enterprises, like cinemas or bus companies, that similarly exemplified modern technology. But newfound affluence was also directed toward more conventional objects. Mothay Gangu Raju at Ellore and Suryanarayanamurti Naidu in Cocanada enhanced their reputation and moral standing through acts of philanthropy—donating land for a dispensary, contributing thousands of rupees to a Hindu charity or religious endowment.

As we will see in the following chapter, rice milling attracted a dual critique. This came from those who saw it (not just in India but across monsoon Asia) as a primary factor in the spread of the nutrition-deficiency disease beriberi, and from those, like Gandhi, who argued that mechanical milling robbed women of their customary work and income. But when it was first introduced rice milling and the appealing white, polished grain it produced attracted little opposition from Indian consumers. By the 1940s a large proportion of the rice consumed in India was machine milled rather than hand husked. In Madras Presidency, where milling was particularly prevalent, nearly two-thirds of all rice consumed came from mills. Certainly milled rice attracted some protests from those who believed the flavor was inferior to that of hand-husked rice (the same complaint was used against mechanical sugar presses). This, however, did little to stem the wider popularity of milled rice. Rice, always a high-status food in India, was now more cheaply available than ever before and could be enjoyed, especially in modern sectors of the economy, among factory hands and plantation workers, even by low-caste groups previously accustomed to eating millets and other coarse grains.

Despite its poor nutritional value, milled rice was preferred by the bulk of the population. It was "more fashionable" and (like the mechanical milling of wheat flour and *dal*) the machine removed "the tedious and time-wasting" task of husking rice by hand.[55] Milled rice exemplified the modern. Attempts to revive the production and sale of hand-husked rice on patriotic or nutritional grounds in the 1930s met with scant success. The physical toil and drudgery of pounding and cleaning rice at home (essentially women's work) was replaced by taking a sack of paddy to the nearest mill (a short walk, bus ride, or cycle trip away). Customers paid a few *annas* a bag to have their rice husked. Alternatively, they surrendered a portion of the cleaned rice to the owner or allowed him to keep the husk and bran—the husk being used for fuel in the mills, the bran for animal feed. For many rice or flour consumers the local mill became, along with bicycles and sewing machines, one of the most familiar machines known to them, an increasingly integral part of modern daily life.

× × × × × × × × × × × × × × × ×

An offshoot of the wider quest for Indian independence and self-sufficiency, the *swadeshi* movement of the early twentieth century was a pioneering campaign to replace British imports with India-made goods. Although the movement found its foremost expression in Bengal between 1905 and 1908, *swadeshi* ideals were far more widespread and enduring, impacting not only on the course and character of Indian nationalism politically but also on, in, and through its economic, social, and cultural manifestations. Despite the setbacks experienced by many early Indian commercial ventures, *swadeshi* became the driving force and prevailing ethos behind nationalist entrepreneurship and consumerism in India. It encompassed commodities as diverse as hand-spun cloth and Ayurvedic medicines, but, as this chapter has shown, it also extended to the sale, assembly, and ultimately the Indian manufacture of bicycles, sewing machines, typewriters, and rice mills. Most commonly identified with manufacturing, *swadeshi* idealism can be further understood as incorporating the process by which even foreign objects were rendered indigenous—even by such simple expedients as giving them Indian names and equipping them with Indian cultural associations.

There were many colonial (and global) constraints on indigenous enterprise, and the Indian response to technological modernity was often quizzical and muted, even where it was not overtly antagonistic. Yet, despite this, Indians found ways to establish their own businesses selling or using foreign machines, or they applied their patriotic idealism and cultural and social values even to modern objects they did not make. This long-term trend toward indigenous appropriation and vernacular use provided one of the bases for the subsequent development of Indian industry and commerce. But, at the same time, cultural ownership engendered or sustained negative as well as positive values. This could, for instance, take the form of denying the use of prestigious goods like bicycles to women or low-caste Hindus. But, taken as a whole, the rise of everyday technology serves to demonstrate how widely disseminated—and how socially and ideologically significant—such seemingly mundane machines had become even before the close of the imperial era.

CHAPTER FIVE

Technology and Well-Being

Machines can be likened to people: they can be understood as serving the needs of human health and well-being. But, conversely, machines can be seen as a force for harm, a cause of individual ill health and a means of mass destruction. It is unsurprising that in India, where responses to technological modernity were mixed and often muted, the ambivalent relationship between humans and machines was central to how everyday technology was perceived and evaluated. As part of the propagandist culture of capitalist consumerism, human health and the body-like functions of machines were repeatedly invoked in the ways in which new technological goods were sold in South Asia as elsewhere. Fueled by the rise of industrial capital and the global quest for mass markets, advertising played a crucial role in the way in which consumer goods were given commercial appeal but also expressed, informed, and ultimately transformed social ideas and cultural values. In an imperial context advertising was, moreover, a powerful disseminator of ideas of race, place, and gender, of desire and exoticism, of civilization and modernity.[1]

Small-scale machines were no exception to this trend. However, as we have already begun to see, propagandist images and consumerist appeals did not pass uncontested. Indeed, in contrast to advertisers' images of health, pleasure, and profit, new technologies encountered criticism as materially harmful and morally corrupting. Throughout the world the alleged human benefits of modern machines attracted skepticism and dissent, but in India, where modern technology had troubling associations with colonialism, where social as well as political argument were often ranged against the machine and its impact on everyday life, this oppositional view held considerable authority.

The very everydayness of the small machine made it a particular target for attack.

Selling the Modern Machine

In an age of growing consumerism and technological change in India from the 1880s onward, the association of machines with images of health was one of the main means by which new technological goods were advertised and sold to the public. There was little specifically Indian or colonial about this: it reflected the broad ambitions of international capitalism and the extension into South Asia of global marketing strategies already deployed elsewhere. Even so, the success of advertising was itself reliant on new technologies of communication, and in this a vital role was played by India's emerging print culture. Barely present in India before the 1780s, by the early twentieth century the printing press had become one of the most widely disseminated of all modern technologies. As well as the dozens of small printing houses in cities like Calcutta, presses were also to be found in most district towns and commercial centers. In 1925 printing presses in India employed 34,000 people, fewer than rice mills (59,000) or tea factories (51,700) but more than jute presses (31,500), dockyards (23,000), and sawmills (16,000).[2]

Sales of newspapers, one of the main products of the presses, were necessarily restricted in a society where levels of literacy were low for men and stood still lower for women. In 1890 there were around 450 newspapers and 100 journals published in India with a combined circulation of 250,000. By 1905 the number of newspapers had risen to around 1,360.[3] Provincial data further suggest the extent of India's print culture. In 1931 Madras Presidency had a population close to 43 million. Of the leading vernaculars, just over a hundred newspapers were printed in Tamil with a circulation of 211,316; roughly half that number of Telugu papers was published, but with sales only about a third those for Tamil newspapers. The province's 58 English-language newspapers had a combined circulation of 93,700.[4] But to the relatively small number of papers sold must be added the ad-

FIGURE 5.1. A Hindi advertisement for Atlas bicycles painted on a north India village wall in 2008. Such advertisements are common in India as a means of selling bicycles and other consumer goods. Author's photo.

ditional individuals who pored over them in libraries and tea shops or heard their contents read out and discussed at village meetings. Newspapers were a visual as well as textual medium, further helping to disseminate awareness of modern goods and machines. Advertisements and cartoons appeared, with simple drawings showing a bicycle, gramophone, or electric fan. By the 1920s photographs of automobiles and airplanes commonly illustrated news items and features. Along with street hoardings, newspapers were one of the most accessible means by which modern goods were made familiar to a wider public.

The power of advertising is evident from a variety of sources. In one of Anand's short stories from the 1930s the narrator, a machine enthusiast, tries to persuade an old-fashioned cobbler to buy a stitching machine by showing him a poster, torn from the walls of the local railroad station. This depicts "an Englishwoman with a bun on the top of her head, wielding a Singer sewing machine." The picture evokes in the illiterate cobbler "wide-eyed wonder," with ultimately fatal consequences, since it persuades him to buy a machine he can ill afford.[5] In Bandyopadhyaya's novel *Aparajito* 1930s Calcutta is repeatedly represented through streets filled with noisy machines and strident advertising clamoring for attention. As the Durga Puja festival approaches, Apu notices a poster for a *swadeshi* match factory "displayed prominently at almost every street corner." Later, as he travels through Delhi at night by train, he sees on the station platforms "all the familiar advertisements"—for Pears soap, Keating's powder, Halls' distemper, and Lipton's tea. The modern world of goods is thus made visible for all to see.[6]

Although initially advertising in India replicated the idioms and images of the West, by the interwar years, in response to the *swadeshi* movement and the rise of Indian nationalism, even Western companies in India were seeking to couch their advertising in forms, such as the use of Hindu mythological scenes and iconography, that would appeal to Indian consumers.[7] By the 1950s, the range of goods advertised in the press was vast, with cosmetics, soap, medicines, biscuits, fans, cameras, watches, gramophones, automobiles, and typewriters among the items most consistently offered for sale.[8] But, because of the "objectionable" (misleading or sexually explicit) nature of the advertisements, or as purveyors of unnecessary and unaffordable commodities, newspapers and the advertisements they carried attracted frequent criticism. Typically for Gandhi the function of newspapers was to educate the public, not to sell goods. He complained that medical advertisements, the largest single source of newspaper revenues, tempted readers to buy medicinal drugs they did not need, or which might even be harmful, while encouraging them to neglect the real moral and physical reasons for their ill health.[9] None of Gandhi's own newspapers carried advertising, and yet without the daily press neither foreign machines nor *swadeshi* goods would have prospered.[10]

The promotion of health and cleanliness, the prevention or treatment of disease, were among themes commonly used in advertising, often trading on Indian ideas of purity and propriety, such as the sale of soaps free from animal fat.[11] The pleasures and frustrations of modern life were also used to market consumer goods. A picture of an overcrowded tram was intended to remind the weary office worker of the benefits of buying a bicycle, a housewife needed a modern stove to please her husband and free time for their social life, and a baby playing with a toy telephone symbolized a modern child whose continuing health and happiness demanded the right brand of gripe water. As everyday commodities, enjoying intimacies of home and body, machines were depicted as healthy, labor saving, and emancipating, a source of sustained well-being to all who bought and used them. Machines appeared as the mechanical companions of modern life; their near-human qualities leant them an almost human corporality. They were ascribed a long and active life, even if it was tacitly admitted that they (or at least inferior brands) might eventually grow old, malfunction, and die. An advertisement for a Remington typewriter in 1904 claimed it "never fails and never grows weary." Another declared: "Remington typewriters are long-lived. Many writing machines break down in their youth, but Remingtons have tough constitutions, and, no matter how hard the work they do, they are sure to reach a hale and vigorous old age."[12] This was not a theme favored by imported advertising copy alone. A notice for an Indian firm of typewriter repairers in Bombay urged, "Let us nurse your sick typewriter back to health."[13]

Bicycles, too, had strong associations with health. To encourage the association of machines with freedom and enjoyment, pictures appeared in the Indian press of happy individuals or smiling couples (often in the 1920s still Europeans) cycling contentedly through a sunny landscape, with only minimal, if any, recognition of the hardships of daily life in India. In 1926 Raleigh improbably promised cyclists "cool comfort" even on "the hottest day," with "no effort required."[14] Twenty years on, the swadeshi makers of Hind Cycles adopted a more realistic note, announcing that their bicycle was "specially manufactured to ideally suit the good and bad roads of India—to withstand the heavy monsoons and the bright sunshine of the tropics."[15] As a pleasurable recreation or rewarding sport, cycling was used to promote other goods, too. In the 1930s the Indian Tea Board ran an advertisement in which a group of energetic Indians exclaimed, after a strenuous cycle race, "Now for a cup of tea!"[16]

Although Singer did not spent much on newspaper advertising, relying instead on the quality and reputation of its machines, sewing-machine manufacturers like Pfaff helped to project the image of well-groomed, smiling women, their hair neatly tied up in buns or wearing elegant saris, who worked joyously at their "singing" machines. For Singer, which in the 1890s produced a series of trade cards showing popular (Western) singers, singing was a recurrent publicity theme and a means of associating its machines with pleasure and pride. On the eve of the imperial durbar at Delhi in 1903, N. M. Patell arranged for a huge arch to be erected over the route to be taken by the viceroy and his entourage. The caption, written in illuminated letters two feet high, read, "Singers Sing God Save the King." To Patell's delight the display drew a wry smile from the viceroy as he passed underneath.[17] Selling goods by proclaiming their pleasurable effects or health benefits was a common advertising ploy and not one confined to machines alone. "Passing Show" cigarettes were sold in India promising "joy in every puff." "Protect your health," ran one advertisement. "If you value your health you should certainly smoke Passing Show. . . . Besides giving you greater smoking enjoyment, Passing Show quickly soothes tired nerves and acts as a refreshing tonic."[18] Automobiles and alcohol were made similarly seductive, as nerve soothing, pleasure creating, and health enhancing.

The Healthy Factory

By the 1930s not only were consumer goods being sold through their association with health or in terms of their health-bestowing properties, but the factories in which they were made were also being presented as healthy and health giving—models of a streamlined, sanitized modernity in which the well-being of workers was as well served as that of consumers. This, too, was an international trend, even if one rather slow to reach India.[19] A news item (doubling as an advertisement) appeared in the *Bombay Chronicle* in 1938, on behalf of Pepsodent, the American company whose "peerless" oral hygiene products—toothpaste, toothpowder, and antiseptic—had already "conquered the world." The company proudly displayed pictures of its "miracle factory," located several miles from Chicago and "far away from the dust and grime and smoke of the city." Visitors were said to be "fascinated" by the factory's ultra-modern machinery and "deeply impressed" by the "magnitude and beauty" of the building and the sunlight streaming in through immense glass windows. All this, the story continued, was done for the benefit of customers—to ensure that the product they bought was "hygienic" and had minimal contact with human hands.[20]

This was not just a distant dream. In June 1938, soon after the Pepsodent advertisement appeared, Lever Brothers (India) announced that it had just opened "India's largest and most modern soap factory" near Bombay. Despite issuing from a foreign firm, the announcement appealed to *swadeshi* idealism, declaring that "Sunlight Soap," made in India by Indian labor, used only pure vegetable oils.[21] Part Western, part homegrown, the social idealism of the sanitized factory, operated by healthy workers, immune to labor unrest, and set in a park-like environment became powerfully established in India between the 1930s and 1960s, at precisely the time when industrial capitalism was threatened by increased worker militancy. The "garden factory" was conceived in eloquent contrast to the crowded, dirty, unhygienic factories that had earlier gained notoriety in India as in the West. This industrial idealism was particularly marked among Parsi entrepreneurs, evolving out of the community's established philanthropic tradition. It followed the example of J. N. Tata whose pioneering iron and steel works at Jamshedpur in eastern India were one of the leading manifestations of *swadeshi* capitalism. But especially significant, from the perspective of everyday technology, was the Parsi firm of Godrej and Boyce in Bombay. Founded by Ardeshir Godrej in 1897, the company began by repairing surgical instruments. Failing to make inroads into that highly competitive market, Godrej turned instead to making *swadeshi* soap. Capitalizing on this success, in the 1950s the firm launched "the People's Soap" which reputedly made it "a pleasure for the common man to keep himself and his family clean and healthy."[22] The firm also expanded into locks, trunks, office furniture, and ultimately, as we have seen, typewriters.

The first Godrej factory was located in congested Bombay, but by 1943 the new head of the firm, Pirojsha Godrej, had developed (in the words of the company's official historian) "a deep concern as a humanist with the problem of industrial slums and the degradation to which they could lead." He decided to move manufacturing to Vikhroli, "an industrial garden township" situated to the northeast of the city, with spacious lawns and well-tended flower beds, in the belief that "working in a modern well-laid-out factory in pleasant surroundings . . . would make work less arduous and more rewarding."[23] After independence the firm instituted an employees' medical benefit scheme and established at Vikhroli a nursery, crèche, and welfare center for workers and their families. It provided dental treatment and a gymnasium, organized yoga and literacy classes, and supported immunization campaigns against smallpox, polio, and tuberculosis. Having become convinced that population control was vital to India's future development, in 1957 the company opened a family-planning clinic, offering women contraception and sterilization. Godrej took the connection between health and technology seriously, and not just in pursuit of worker satisfaction (there were no strikes at Vikhroli until 1972) and the rewards of industrial paternalism. One of the firm's sales strategies in the 1950s was to give typists a free gift of Godrej soap and toiletries with each new typewriter sold, in part to overcome office workers' resistance to its new machines.

Industrial idealism and the desire to create "healthy relations" between employers and workers were also evident in the opening in 1952 of a factory for the Sen-Raleigh Company at Kanyapur in West Bengal, to mass-produce bicycles for the Indian market. Sudhir Kumar Sen's patriotic objectives were realized not only through a collaboration with Raleigh of Nottingham that brought the latest bicycle-making expertise to India but also by the creation of an ultra-modern factory, set among flower beds and tree-lined avenues. Equipped with "rows of precision coin-

FIGURE 5.2. A cycle assembly line at the Sen-Raleigh plant at Kanyapur, West Bengal, in the early 1950s. Reproduced from the Raleigh Archives (DD/RN 6/8/3) by kind permission of the Nottinghamshire Archives, Nottingham, UK.

ing presses, giant hardening furnaces and automatic assembly lines," the factory boasted huge, moveable windows, designed to maximize the availability of natural sunlight and fresh-air ventilation, both of which were said to be "essential aids to production" in the Indian climate.[24] The township that grew up around the factory had its own medical and recreational facilities and was further intended to exemplify the employers' commitment to industrial health and worker welfare. According to publicity statements, the Kanyapur factory, expressly "planned for modern India," would "serve as a blueprint for other factories in the country."[25]

Disease and the Machine

It might be asked whether the correlation advertising made between companionable machines and healthy, happy lives actually succeeded in shaping public attitudes and consumer behavior. In some ways it appears that it did. Ironically, not least in view of Singer's rampant commercialism and its success as one of the harbingers of international capitalism, Gandhi bought into the image of the felicitous sewing machine, believing that Isaac Singer, the firm's founder, had created the machine as a gesture of love for his wife and a means of saving her unnecessary toil. Seemingly made for love, not profit, the sewing machine was one of the few modern technologies to which Gandhi gave his approval.[26]

A further, more curious, example appears in the memoirs of Prakash Tandon, whose father was an engineer working for the Punjab irrigation department. In 1918 Tandon senior fell seriously ill with influenza in the pandemic that cost India an estimated twelve million lives. It took him a long time to recover and in order to cheer himself up he decided to buy a Singer sewing machine. (Why not a gramophone? Who in the Tandon household was going to use a sewing machine?). Although the Tandon family had many connections with modern technology through their father's work, the sewing machine, "shining black and chromium-plated," was the "first mechanical contraption" to actually enter their household. Attracted by the company's propaganda about singing machines and by the colorful calendars it issued with its machines, Tandon believed the Singer was a happy machine and exemplified "the international love of singing." In buying a sewing machine he also encouraged his children to learn to sing.[27] In India, as in many other parts of the world, singing was closely identified with health and healing. One of the ways in which Hindu women tried to ward off the destructive effects of smallpox was to sing songs in praise of Sitala, the goddess believed both to bring the disease and to control its severity. Tandon seemed to be following a musical tradition that linked singing, in this case via the machine, with well-being.

Perceptions of disease and the machine—or its users—could be combined in a variety of ways. For instance, as intermediaries between the European household and the unsanitary bazaar, *darzis* (despite their modern machines) were suspected of transmitting smallpox, cholera, and other diseases.[28] Equally, woven into the everyday narrative of commerce, a larger history of disease loomed. The period 1890 to 1920 marked "a woeful crescendo of death" in India, with cholera, plague, and influenza among the most deadly diseases.[29] From the mortality and social dislocation it caused, the plague epidemic that began in western India in 1896 created particularly testing conditions for the dissemination of everyday technology. Large numbers of workers quit Bombay, leaving city streets and mills deserted. The disease and, still more, British attempts to contain it impacted widely on the countryside. Postmen and other state employees died from their occupational forays into infected areas. Exposed to plague-carrying fleas hidden in old clothes, tailors were particularly at risk of infection. As Patell moved through western and central India in 1902–1903 he found few tailors to sell or loan sewing machines to: they had all died or left for famine relief works. One of Singer's own salesmen died of plague, and Patell's efforts to travel through the stricken region were hampered by antiplague cordons that barred his entry into towns and districts or required him to report to medical officers for inspection and certification. In writing to New York, Patell tried to impress upon his bosses how dire the situation in India was and how meager the prospect of selling machines to a hungry, disease-ravaged population was.[30]

However, one could read the association between health and the modern machine as an essentially positive one. By the interwar years, mobile dispensaries were penetrating the Indian countryside, furthering the spread of vaccination and providing basic medical facilities. "Magic lantern" shows were widely in use for health propaganda. Cycle races for both men and women demonstrated physical fitness as well as an appetite for speed and competitiveness (though they might also occasion death and injury to exhausted, and perhaps undernourished, competitors).[31]

Modernity and Its Discontents

Clearly, in India, as elsewhere, many individuals delighted in the modern machine, reveling in its speed, noise, and stylishness.

As a child in Punjab, Prakash Tandon was fascinated by machines. Already intrigued by the instruments and gadgets in his father's office, when the telephone arrived his "excitement knew no bounds."[32] By the 1930s Indians, or at least the more affluent among them, were being invited to take aviation lessons and "see India from the air," or to join the growing "radio habit."[33] Some welcomed the coming of the modern machine as a means to advance social reform, or to improve the well-being of gender, class, and nation. Others, more pragmatically, simply accepted it as something to be lived with or as a necessary adjunct to their work.

Historically, the rise of rampant consumerism has often tended to provoke adverse reactions and create discomfort at the crass and amoral ascendency of the "world of goods."[34] In India, where mass poverty and hunger prevailed and famine was no distant memory, there seemed to many something repugnant about the "craze" for consumer goods and the affluence and indifference that accompanied them. Although Gandhi is the individual most commonly seen as the opponent of India's "modern civilization," his was not the only voice of dissent, nor yet the earliest. To a degree possibly unparalleled in other "modernizing" societies in Asia, the critique of technological modernity in India tapped into a powerful mood of anticolonialism, one that identified machines with the arrogance and exploitation of imperial rule.[35] More than that, it connected with a deep vein of social conservatism and intellectual pessimism about the alleged benefits of modern, machine-driven progress. Nostalgia for a preindustrial age might be an inescapable part of modernity itself, but in India this critique was augmented by feelings of powerlessness, frustration, and rage. At the height of the *swadeshi* movement in 1906, Aurobindo Ghosh likened the influence of the West to the spread of tuberculosis and called upon Indians to spurn "like deadly poison all those misnamed ideals so dear to the West"— industrialism, consumerism, and imperialism.[36] Modernity, thus construed, was the very antithesis of health.

Among high-caste Hindus, particularly in Bengal, regret for the loss of an imagined past age of health and prosperity merged with a growing sense that they were losing out not just to the callous, greedy British but also to the more enterprising Marwaris from Rajasthan or the fecund, meat-eating Muslims, and were in danger of becoming, in the Darwinian language of the time, a "dying race." Deteriorating health was one theme of this critique, the "machine craze" another. A leading critic was Pramatha Nath Bose. Previously an advocate for Western technical education in India, by 1901 Bose had come to see modern technology as an imposition that facilitated Britain's economic exploitation of India and an illustration of the domination of the "unscientific peoples of the world" by the "scientific peoples of the West."[37] The mechanized warfare of 1914–1918 further persuaded him of the destructive nature of the so-called civilization that had taken hold of India. With Gandhi's emergence at the end of the war, Bose was more than ever convinced that Western technology served only to deepen India's subjugation and hasten its "physical decay and degeneration." Once-honored ideals of simple living and renunciation had perished in the "frenzied race for wealth and luxury." Although in Bose's view India was not the only country affected ("the whole world" had become a "scene of destitution, disease, vice and malevolence"), Hindus faced extinction "at no distant date" if they persisted in being the "imbecile imitators" of an "immature civilization." Modern machines served only to accelerate physical degeneration and moral decay.[38]

Although at the extreme end of the spectrum of antimodernity, Bose's views were not without parallel in the speeches and writings of his fellow Bengali, the pioneering chemist Prafulla Chandra Ray, and, more influentially of course, of Mohandas Gandhi. Significantly, from the viewpoint of this discussion, everyday technology was as important to these critics as big technologies like the railroad. Indeed, in some ways the everyday machine was even more significant, for it impinged more directly on the daily lives of the people and on the life of the village, the heart and soul of the nation. Ray inveighed against the delusion that led peasants to buy bicycles and other goods they could ill-afford and had no real need for. But one of the principal targets of his critique (and that of many contemporaries) was the way in which mechanization was changing traditional food processing and consumption. In a pattern common to many other parts of the world where men with machines replaced women in food processing, in India this included the usurping of female labor in husking and cleaning rice and in grinding wheat and pulses.[39] Not only did mechanization rob poor women of much-needed employment and income. It also produced foodstuffs that were said to be inferior in quality, taste, and nutrition to those it displaced.

Since rice could only be consumed after the inedible outer husk had been removed from harvested paddy, a great deal of arduous manual labor was required in its preparation. It was estimated to take an hour's pounding a day by two women to produce enough clean rice (six to eight pounds) to feed five people.[40] In a few areas rice husking was done by men, but across most of the country it was overwhelmingly women's work. At least half a million women were said to engage in this occupation in 1901, making it one of the commonest forms of female labor, even ignoring the many women who husked rice as part of their daily domestic routine.[41] In some areas women from two or three households worked together husking, keeping time with their pestles and with each other by singing "pounding songs." These songs, "the most beloved in Indian folklore," disappeared as the repetitive drone of the milling machine took over.[42]

Customary husking practices varied from one part of India to another. One common technique involved the use of a wooden pestle, about four feet long. As this was thrust into a wooden or stone mortar containing the raw paddy it dislodged the outer husks. The separated grain and husks then had to be scooped out by hand.[43] The other common type of husking, particularly in Bengal, involved a more elaborate wooden apparatus known as a *dhenki*. This "indispensable domestic utensil" consisted of a mortar, "excavated out of a log of wood, and . . . sunk in the ground," and a pivoted wooden beam to one end of which was attached a wooden pestle that rose and fell as the foot pedal at the other end was depressed and released. The *dhenki* needed the labor of two women: one to operate the pedal, the other to feed paddy into the mortar and extract the husked grain.[44] The *dhenki* was

FIGURE 5.3. A south Indian woman uses a pestle and mortar to husk paddy, with a winnowing fan at her feet. Late nineteenth-century photograph, photographer unknown, image S0002117, reproduced by kind permission of the Royal Geographical Society, London.

more than a utilitarian tool. It symbolized domestic harmony and orderliness of household life. The thud-thud of the *dhenki* was a sound women grew up with; hearing and working the machine became a cherished childhood memory. In high-caste Bengal it bore additional ceremonial significance as women used

FIGURE 5.4. A foot-operated rice-husking machine, similar to an Indian *dhen-ki*, on display at the Vietnam Museum of Ethnology, Hanoi, in 2008, a reminder of how, until recently, many of the technologies of rice production and process-ing were common to South and Southeast Asia. Author's photo.

the *dhenki* to prepare ritual rice for household worship.[45] In the novel *Pather Panchali* by Bibhutibhushan Bandyopadhyaya the girl, Durga, yearns to see the train she has only heard in the dis-tance. By contrast an ancient aunt living with the family recalls the old days when the dutiful wife pounded rice by hand: "She could hear the noise of the husker, and see her [the wife's] golden bangles sliding up and down her arms as she worked."[46] If the train was a cogent symbol of technological modernity, the *dhenki* was a potent icon of domestic tradition.

Dramatic changes followed the arrival of the modern rice mill. Its first (and arguably greatest) was with respect to women's work. As early as 1908 it was noted in Bengal that milling was having a deleterious impact on women from poor families who had previously earned a meager subsistence from husking and cleaning rice and for whom there was now little alternative em-ployment.[47] The 1911 census reported a substantial decline in this form of women's work, and within two decades hand husking had become relatively rare, especially in south India. In Madras Presidency alone the number of women whose occupation was given as rice pounding fell from 56,622 in 1921 to 38,701 in 1931.[48] In the 1930s, K. Ramiah noted that traditional rice husking was now "practically extinct" except in some interior parts of the west coast districts of Malabar and South Kanara, adding, "Because of the hard and tiresome nature of the hand-pounding and due to the rapid increase in the number of power-driven hullers even in the rural parts of the country, it has become difficult to get labour for pounding rice. Even the coolie classes who get wages paid to them in kind, usually take it to the nearest mill to get it pounded."[49] In a radical reversal of earlier work practices, the labor in the new mills was largely male, especially in skilled jobs such as operating machinery. Women were employed in smaller numbers, as casual or seasonal workers, to perform menial tasks like raking the parboiled grain to dry outdoors in the sun. They received lower wages than men, only marginally above the rates paid to female "field coolies."

The Ills of Rice Milling

The argument against mechanized rice milling came from several quarters. It was partly made by British observers who, preferring a rural idyll to a semi-industrial landscape, regarded the modern machine as an alien and unwelcome presence in village India, destroying the tranquility and self-sufficiency of rural life. But it drew still more influential support from the growing body of medical evidence that linked mechanized milling with the spread of the nutrition-deficiency disease beriberi. Where the crudeness of hand pounding left much of the inner skin of the rice grain—the pericarp—intact and so retained the vitamin B_1, or thiamine, it contained, mechanical milling and polishing stripped rice of virtually all its vitamin content. Instead of being used for human nutrition, the bran was relegated to feed animals. For people who lived almost entirely on rice, the absence of thiamine could result in painful debility, stiffened limbs, and eventual death.[50] By the 1930s an extensive medical literature had developed maintaining that the rapid rise of this disease across monsoon Asia, from India to Japan, and the mortality resulting from it, was directly attributable to milling. From its inception in 1910, the Far Eastern Association of Tropical Medicine, meeting annually at cities across East, Southeast, and South Asia, passed resolutions making explicit the connection between milled rice and beriberi, and pointing the finger of blame at the ever more popular rice mill and its product. Some medical and nutritional experts wanted milled rice banned or taxed to deter consumption. Less drastic alternatives, such as using an extract of rice bran as a food supplement, were also proposed, alongside propaganda to encourage balanced diets.[51]

Medical and administrative opinion in British India was not entirely supportive of this view of the "beriberi problem." Still wedded to the doctrine of laissez-faire and mindful of the popularity of white rice and the economic power of grain merchants and millers, the government was loathe to ban white rice or check the proliferation of rice mills. Although beriberi had been recorded in India since the early nineteenth century, it had not spread as rapidly there as in many parts of Southeast and East Asia: many Indians consumed rice that had been parboiled before husking, a process that preserved part of the thiamine content. Beriberi was, in fact, only prevalent in one part of India—the Andhra delta—where exceptional conditions existed. However, pioneering investigations, first by Robert McCarrison of the Indian Medical Service and then by W. R. Aykroyd, McCarrison's successor at the Nutrition Research Laboratories at Coonoor in south India, demonstrated the nutritional poverty of rice compared to other food grains like wheat and millet and used beriberi to illustrate the wider problem of endemic malnutrition in India.[52]

Even if the beriberi problem appeared less acute in India than in other parts of monsoon Asia, its existence there underscored the extent to which a shared technology—the modern rice mill—posed a common threat to public health, especially where poverty and deficient diets remained widespread. Medical expertise provided evidence both of the spread of the machine and of its deleterious effect. As W. R. Aykroyd and B. G. Krishnan observed in 1941 of paddy-producing tracts in the Godavari delta, the rice mill had become "a prominent feature of the landscape." "Driving through the district one observes large mills at intervals of 10 miles or thereabout, and even in comparatively remote villages small mills . . . are to be found."[53] Their position was clear: "While living standards in India remain at the present level, we must view with alarm any extension of mechanical rice milling in rural areas."[54]

Given the force of the medical arguments against rice milling it is not surprising that its links with mechanized milling became established, too, in the Indian critique of technological modernity. In 1921 Bose cited beriberi as one of the adverse consequences of technological change and the abandonment of a diet that, honed through "centuries of experiment," had produced Indians whose exemplary health was reflected in their "splendid physique and . . . mental vigour."[55] Gandhi, too, used the medical evidence of McCarrison and Aykroyd to lend scientific support to his arguments against mechanized milling and for a return to traditional rice husking and other village occupations. In 1934 he called for the replacement of sugar mills and their produce with traditional sugar presses and the "vitamin-laden and nourishing *gur* or molasses" they produced, just as he favored unpolished rice "whose pericarp, which holds the vitamins, is left intact by these [hand] pounders."[56] McCarrison and Aykroyd did not entirely accept Gandhi's interpretation of their findings, arguing that the adoption of a more varied diet could adequately compensate for the thiamine deficiency in white rice and that less rigorous milling would preserve enough vitamins to prevent beriberi.[57]

But the argument made by Gandhi and P. C. Ray was more wide ranging than nutrition alone. They advocated a return to hand husking, believing that it, through the physical labor involved, was conducive to women's health, that it gave employment to needy women, especially widows, and that it fostered the self-sufficiency of the Indian village community. Gandhi argued that rice and flour mills together had "put thousands of women out of work and rob[bed] them of [their] health."[58] To him it was clear that "human greed" was "responsible for the hideous rice-mills one sees in all the rice-producing tracts." He urged the public to appeal to the mill owners "to stop a traffic that undermines the health of a whole nation and robs the poor people of an honest means of livelihood."[59] For his part Ray observed that paddy husking had hitherto been "the only home industry in Bengal by means of which poor widows with infants in arms could eke out a miserable living." Almost every household had had its *dhenki*. Now, thanks to the so-called "march of civilization," rice mills were "springing up all over Bengal," depriving poor women of an income. "When it is remembered that a single rice mill snatches away morsels of bread from the mouths of hundreds of the destitute, the result can well be imagined. A few capitalists are lining their pockets at the expense of their helpless sisters."[60]

There was support for these views in a detailed investigation published in 1934 by Hashim Amir Ali of Vishva Bharati University into the growth of rice mills in the neighboring town of Bolpur. This showed that since the mills' inception in 1913 an estimated 8,000 women had lost their domestic employment (and hence their income) from rice husking to the roughly 1,350 (mostly male and partly seasonal) workers the mills now employed. In addition, the mills spread smoke pollution over a wide area and deprived cattle of the husks on which they had previously fed.[61] And yet, while the loss of female employment in rice husking was widely deplored, criticism was not unqualified. Many women (or the scholars who wrote on their behalf) were relieved to be free of the arduous daily task of pounding rice or grinding flour in a hand-turned quern. A Hemlata Dandekar put it, speaking for her women informants in rural Maharashtra in the 1960s, the grinding of cereals and pulses by hand was "a backbreaking task which has now, thankfully, been eliminated."[62] Men were more likely than women to argue that the physical toil of rice pounding was good for women's health. Equating women's work with hard labor and urging the benefits of traditional technology has often been a way of denying women the advantages of modern technology and of binding them still closer to the traditional social order.[63]

In the Factory

The onset of technological modernity and the impact of the machine age on human health provoked a wide-ranging debate in late colonial India. This took many different forms and represented varied standpoints, not just those of the social conservatives and Gandhians. Some of Mulk Raj Anand's stories, written from a leftist perspective, suggest the potentiality of the machine as a vehicle for the social liberation of India's most oppressed communities. At the close of his novel *Untouchable*, Bakha, a sweeper, learns of the flush toilet, a "machine which clears dung without anyone having to handle it" and of its promise of a casteless, classless society.[64] But Anand's work also shows a far more visceral hostility toward machines as instruments of capitalist exploitation, and here it is the violence of modern technology that prevails, not its potential for social liberation. In some of his stories the machine, encountered on the street or in the factory, is likened to a predatory beast that, even when it does not kill or maim outright, sucks the lifeblood out of the poor, suffering worker. Trucks, automobiles, and bicycles constantly threaten to run down the rural innocent who has strayed, hungry and in search of work, onto unfamiliar and bewildering city streets.

But the oppression of the machine is also evident in the countryside and in the many makeshift factories in which agricultural and industrial commodities are first processed. In one particularly brutal short story by Anand, entitled "Lullaby," a young woman sits working amidst the "thick, sickly, tasteless air" of a jute-processing shed. Fluff flies everywhere, but, as the woman is kept constantly busy "feeding the gaping mouth of the machine," she has no time to feed the hungry child in her lap. Noise is everywhere in this tale—the noise of the engine and conveyor belt, the bolts jumping up and down in their sockets—drowning out the woman's soothing lullaby and the child's feeble cries. In the end the child dies but the "jaws of the monster" continue to demand to be fed.[65] In another story, referred to previously, a cobbler is encouraged by his young friend to share in their "common ... passion for the machine." He invests in a stitching machine, so that he can produce shoes more quickly and cheaply. However, in order to acquire the machine he has to borrow heavily; this lands him deeply in debt, and he has to work harder and harder to repay his loan. The effort is so great that the cobbler, "drained of his life-blood" by the machine, dies, "killed by the devil disguised in the image of the sewing machine."[66] Finally, in *Coolie*, published in 1936, Munoo is a youth from the Punjab hills who experiences both the novelty and fascination of the machines (like hearing a gramophone for the first time) and their bestial nature. He goes to work in the Bombay cotton mills, where the deafening, unguarded machinery, like a "many-headed, many-armed machine god," is a constant danger to life and limb, and where choking dust and fiber fill the hot, unventilated room. Ultimately, he contracts tuberculosis and, though he quits the factory to return to the hills, he dies of the disease.[67] The message of all these stories is clear: under capitalism, modern machines are instruments of exploitation and destroy the lives of those who work them.

These were, of course, expressions of Anand's literary imaginings and his middle-class understanding of the subaltern experience of technological modernity, but he had ample contemporary evidence to draw on. His views reflect the social reality of the time and the largely unchecked capitalist exploitation of industrial workers and the urban poor. The experience of everyday technology for those who worked with machines or came into daily contact with them was strikingly different from the images of health with which this chapter began.

Studies of factory labor in India have tended to focus on the larger, more organized industries, such as the jute industry in Calcutta, sometimes touching on their impact on workers' health.[68] Far less attention has been given to smaller factories and the conditions of their workers. In part this reflects the dispersed nature of these industries and the inability of the state to observe them as closely as it could larger factories in Calcutta or Bombay. But it also suggests relative unconcern on the part of the Indian public and emerging trade unions. One indication of the violent career of the machine and its consequences for workers' health can be given by returning to our example of the rice mills.

Although the critique of rice mills was primarily directed at the labor they displaced and the nature of the rice they produced, that critique might equally have been directed at what happened to the health and well-being of those who worked in the mills. In contrast to the auspicious names given to them by their owners, rice mills became for those who worked in them places of injury and, not infrequently, death. The rise of the rice mill coincided with the growth of factory legislation in India, though until 1911 many rice mills were too small to be inspected.[69] Thereafter rice mills were among the places most commonly reported for serious accidents and violations of the factory acts. "Loose" clothing (such as dhotis, saris, turbans, and shawls) was caught in whirling machines, dragging workers to their deaths; women's hair was trapped in conveyor belts and flywheels; internal walls and grain stores collapsed, killing "coolies" or causing broken ribs and fractured limbs; men, women, and juvenile workers carrying heavy loads of grain tripped and fell into vats of boiling rice. In trying to repair or clean machines (many devoid of safety guards) or to remove waste from under moving parts, workers lost fingers, arms, and legs, or contracted tetanus and died a few days later. Due to faulty wiring, they were electrocuted by faulty machines and light fittings.[70]

That much of the work in rice mills was performed by casual, unskilled, often seasonal labor and was carried out in small, largely unregulated premises, some of which were "totally unsuitable" for industrial use, added to the hazards of the workplace.[71] Fires were frequent in both rice and flour mills.[72] The quest for quick profits and the ready availability of cheap labor made factory owners and managers still more negligent of workers' well-being. Until the late 1930s few even of the larger mills in India had on-site medical and sanitary facilities or canteens where workers could eat and rest between shifts. Considering the size of the workforce and the nature of the conditions involved,

FIGURE 5.5. Loading paddy into a village rice mill in Tanjore [Thanjavur] in south India in 2008. Author's photo.

it is perhaps surprising that there were not even more deaths (saw mills in Burma were far more dangerous places in which to work). In the decade from 1922 to 1931 there were 35 deaths and 32 serious injuries recorded in Madras Presidency. In the following decade (1932–1941) 10 deaths and 64 serious injuries were reported (though, given lax management practices and inadequate inspection, many accidents passed unrecorded). In Burma, the only other province for which detailed statistics were given for rice mills, in the period from 1932 to 1940, there were 49 deaths and 136 serious injuries.[73] Because of the rural location of many rice mills and similar small factories, observers were inclined to regard them as relatively healthy places to work. Long-term injury to workers' health, due in particular to the constant noise and the prevailing "dust evil" produced by the milling process, was also largely ignored in inspection reports. Where, as in the cotton-ginning factories of Punjab, the dust problem was highlighted, the inspectors felt themselves powerless to prevent it: the costs of installing adequate ventilation would be prohibitive for such small enterprises.[74]

While lax management practices received their share of the blame, it is striking how often in these factory narratives workers were blamed for their own injuries and fatalities. If one of the expectations of the modern machine was that those who used it—whether in the home, the street, the factory, or the penal institution—should obey the discipline of modern mechanized life, then one of the criticisms levied at many working-class Indians was that they failed, even at the cost of their own lives, to meet those corporeal and behavioral norms. The Madras labor commissioner remarked in 1920, following a year in which 7 deaths, 6 serious injuries, and 656 minor injuries had occurred in the province's rice mills, that the majority of accidents were not the managers' fault. Rather, they were "entirely due to the carelessness and stupidity of the operatives. They may almost be said to go out of their way to court disaster and this is undoubtedly due to their lack of intelligence or common-sense and at times even to inquisitiveness as to the mechanism of certain machinery which fascinates them. Month after month, year after year, the doctrine of care and self-preservation is propounded to the workmen but results are far from satisfactory."[75] Mill owners and managers were fined under the factory acts (indeed they were among the most common recipient of penalties under the acts), but often the fines were too small to deter future malpractice or incentivize improvement. A rice mill at Dinajpur in Bengal in 1934 was adjudged so dangerous by a factory inspector, with "imminent danger to life and safety," that it was ordered to cease working until changes were made. When the manager ignored the order, a fatal accident resulted. But, despite this flagrant defiance of the factory legislation and the institution of a formal legal case against him, the manager was only fined Rs 100, "an obviously inadequate amount," in the words of the provincial chief inspector of factories.[76] Besides, many accidents went unreported (with employees having little reason of their own to report them if they wished to remain at work), and the small staff of factory inspectors was too inadequate to carry out more than occasional checks. At a time when laborers in textile mills and on the railroads were becoming unionized, the often seasonal and casually employed workers in the small and scattered rice mills seldom belonged to unions or had the organizational capacity to strike for better wages and working conditions. Workers were reluctant to complain about their pay and conditions for fear of losing their jobs.

× × × × × × × × × × × × × × × ×

Modern advertising projected, not unsuccessfully, the idea of modern machines and consumer goods as companionable objects, conducive to human health and well-being. Further, modern goods might be produced in modern factories where the health of both producers and consumers was duly accommodated. Ranged against this positive view of the modern machine was an antithetical set of views that contested the healthiness of the machine and machine-made goods and, in a spirit of cultural nationalism, extolled the superior virtues of a preindustrial age of production and consumption. In India the critique of technological modernity took many different forms, but it drew in particular on the strength of association between colonialism and exploitation, and the conviction that a poor and populous country like India had no need for the labor-saving luxuries and affluent indulgences of the West. If this critique in some ways reflected the raw subaltern experience of the factory, or other everyday examples of machine use, in others it appeared to mirror colonialist claims that India was unprepared for the trials and opportunities of modern, mechanized life. In field, factory, or street Indian bodies were, as far as mechanized modernity was concerned, all too often bodies out of place and out of time, ill-equipped to face the demands and challenges modernity posed. But it is notable, too, that the critique of technological modernity extended to small machines, like the rice mill, and was not con-fined to large industrial undertakings and expensive objects, like automobiles and airplanes. Indeed, the small machine seemed to occupy a place of particular prominence.

CHAPTER SIX

Everyday Technology and the Modern State

Modern technology and the modern state grew up together, the one lending strength to the other and fueling their mutual ambitions. But state engagement with technology took many forms. In the Indian context, the colonial state has often been criticized—justifiably—for its relative absence. It is accused of having done too little to assist the growth of indigenous industry and to promote the dissemination of modern technology, except in so far as these facilitated Britain's control over India and its material and human resources.[1] As a colonial power, the British are seen, despite occasional spurts of activity, as indifferent, even openly hostile, to the growth even of basic industries, and preceding chapters have given some demonstration of how this was so. And yet the state can be viewed in a more varied light than one simply of self-interest and neglect. From about 1900 onward, India's colonial state was evolving rapidly into a modern state, and one of the ways in which this modernity was expressed was through modern technology and the exercise of mechanical power. To a degree unmatched in Western societies (where commerce, industry, and civil society played a more dominant role in fashioning technological modernity), the colonial state in India was a leading user of, and publicist for, modern machines, including those with which this book is concerned.

The colonial state in India might often appear to be alien, aloof, and preoccupied with grand engineering projects. And yet it employed tens of thousands of subordinates whose working lives brought them into close contact with small machines or whose daily activities introduced machines into the everyday lives of others. The subaltern existence of the machine was thus to be found within the apparatus of the state as well as in the wider ranks of society. The state might have used machines to make—or try to make—compliant, disciplined subjects, but also found itself drawn into their routine governance. Further, technology was implicated in the rise of "the everyday state." This phrase is conventionally used to describe the ways in which the postcolonial state, or its local and subordinate agencies, came into frequent contact with the people, obliging them to interact with it, often in ways (as through nepotism and corruption) remote from formal state structures.[2] In fact, the kinds of everyday technology discussed here also played a formative part in this process of engagement, negotiation, and negation, and with the colonial, as well as postcolonial, state.

The challenge of Indian nationalism and left-wing movements, the growth of urbanization, the rise of factory production and an industrial workforce, and the emergence of a more interventionist approach to public health and labor control—these factors helped impel an often reluctant state toward greater engagement with the society over which it presided. Hastened by the impact of two world wars, these changes helped force the eventual abandonment of the laissez-faire ideology that had dominated nineteenth-century colonialism. Yet even in the heyday of "liberal" empire, the British sought to control vital areas of technological activity—such as the construction and management of railroads and irrigation works—or to intervene where technologies like the printing press seemed to threaten the security of the colonial order. Economic and social change, themselves linked to technological innovation, demanded new regulatory mechanisms, as in the case of factory legislation or measures to control the cinema and radio. But this did not simply result in untrammeled domination by a new, technologically empowered, state. Especially in a colonial situation, where technological resources were limited or were less efficiently deployed than in Western societies, there was the persistent possibility of the state's technological power being undermined by its own weakness or being overwhelmed and appropriated by its opponents. Technology, not least everyday technology, might become a site of resistance and not merely conformity.

Technology and the Inner Life of the State

In the late colonial era machines became increasingly important to the exercise of British power over India. This can, of course, be seen in terms of "big technologies," from railroads and telegraphs to tanks, armored cars, and even aerial surveillance and bombardment. But our concern here is with small-scale technology, and, as with the critique of technological modernity, it was often small machines that mattered quite as much as large technological systems. We will begin by examining the place of such machines in the internal economy of the colonial state—first in relation to transport and then the office.

A physical ability to move around India, through its vast and varied landscape, was always vital to India's colonial regime. District officers were almost constantly on the move—haranguing village headmen, inspecting public works, overseeing settlement operations, observing famine, flood, and pestilence, reporting debt and disorder. For much of the nineteenth century this mobility was mostly achieved by relatively primitive means—on horseback, supplemented by the use of palanquins (enclosed litters carried by four bearers), boats, oxcarts, and occasionally elephants and camels. The coming of the railroads in the 1850s made long-distance travel quicker and easier, and greatly facilitated the movement of troops; but the railroad did little to diminish administrative reliance on touring by horseback. Ability to ride a horse remained an essential requirement of civil servants and police officers well into the twentieth century.

However, the arrival of the internal combustion engine, along with novel devices like the telephone, began to undermine equestrian colonialism. At first automobiles were the sole prerogative of the upper tiers of the colonial hierarchy and special sanction was required for their purchase from official funds. However, small though the number of automobiles in India was even by the 1920s, among the elite—European and Indian alike—they quickly became an established mode of transport. Gilbert Slater related how, early in the century, some Europeans had laughed at the idea of Indians driving automobiles on the familiar grounds that they were by nature too timid and technologically inept.[3] But, to the consternation of skeptical colonialists, Indian owners and drivers were soon the majority. For a while, India's landholding and princely elite took to the recreational use of the bicycle, but, once automobiles arrived, that fashion soon waned.[4] When the Webbs toured India in 1911–1912 they repeatedly encountered Indians with expensive automobiles. By the time Edwin Montagu, secretary of state for India, visited in 1918 virtually every prince owned at least one automobile: the rare maharaja who did not was ridiculed as old fashioned.[5] Indian industrialists, commercial magnates, lawyers, and others from the professional classes rapidly acquired them, too. Motilal Nehru, a leading Allahabad lawyer and father of Jawaharlal Nehru, was famously the first Indian to own an automobile in the city in 1907. Criticized for purchasing a foreign vehicle when Indian thoughts were turning to *swadeshi*, Motilal wrote testily to his son, "Would you advise me to wait till motor cars are manufactured in India[?]"[6]

So widespread had use and ownership of automobiles among affluent lawyers become by 1930 that officials in Madras responded to Indian participation in Gandhi's civil disobedience movement by confiscating their vehicles. When automobiles belonging to pronationalist politicians were wrecked during a police charge, the impression was further created that the British were targeting these machines as a way of attacking wealthy Congress supporters.[7] Among many nationalists the morality of the motorcar remained a matter of debate. Gandhi voiced unease at having to use one.[8] In campaigning for the provincial legislature in Madras in 1936–1937, the veteran Congressman and Gandhian C. Rajagopalachari used the image of current ministers and their supporters riding in motorcars to emphasize their aloofness from the people.[9] And yet when, after the elections, the Congress itself formed ministries in several provinces (including in Madras under Rajagopalachari), traveling in an official car became a prerequisite of office, a visible sign of political empowerment.

It is hardly surprising that officials saw not only the utility of the automobile but also its prestige. If the British, who often saw their authority reflected in the mirror of Indian opinion, weren't to lose out to Indians, it was necessary to keep pace with technological change. The humble bicycle wouldn't serve this purpose. But the automobile, too, posed difficulties. At first, it was argued that if the use of motorcars was allowed for every district officer then the value of their touring, and hence their close contact with the rural population, would soon be lost. If they became reluctant to leave main roads for tracks unfit for motoring, the prestige and visibility of the Raj would quickly diminish.[10] But this argument was replaced by the claim that automobile use actively augmented state power. It saved official time, allowed magistrates and police to move quickly to the scene of a disturbance, and enabled factory inspectors to visit many more locations than would ever be possible by horse or bike. At a time of mounting unrest it was thought inefficient and unseemly for senior police officers to be dependent on horses or the use of shared vehicles.[11]

Between 1905 and 1920, a widening circle of state servants were allowed to purchase automobiles or were given advances to buy one for personal as well as official use. But the process of making rules, vetting applications, and deciding which kind of vehicle should be sanctioned spawned a whole new domain of bureaucratic activity. That a comparable, if far less costly, expansion of bureaucratic activity surrounded loans for the purchase or maintenance of bicycles for low-ranking government employees emphasizes the stark divisions of race, class, and status between the upper and lower echelons of the colonial state in India.[12] It is worth noting, too, that while the market in bicycles favored British manufacturers, it proved impossible to pursue a British-only policy for automobile purchases. French, German, Italian, as well as British, motorcars were bought on grounds of cost, utility, and taste. With the virtual cessation of imports from Britain and Europe during World War I, dominance over the Indian automobile market passed to American manufacturers, a position retained even when peace returned. In 1920–21 barely 2,500 imported automobiles were British-made compared to 10,000 of

American origin.[13] Newspaper advertisements for American cars far exceeded in size and number those for British manufacturers. Despite the prestige enjoyed by carmakers like Rolls-Royce, the majority of automobiles in India in the 1920s were of non-British manufacture, as were many trucks and buses.[14] It is ironic that one British official, explaining to the American journalist Katherine Mayo the advantages motor transport brought to the delivery of famine relief, exclaimed, "God bless Henry Ford!"[15]

Typewriters represent a different aspect of the internal technologization of the colonial state. Earlier chapters have shown their steady, if unspectacular, increase in India from 1900 onward, but the typewriter had a particular significance for colonial office life. Government departments and agencies were among the main purchasers of typewriters and valued customers for maintenance and repair work. Manufacturers strove to obtain government contracts and used state patronage to publicize the merits of their machines.[16] As with automobiles, American machines far outsold British ones. Even by 1910 there were really only two makes of office machines to choose from—Remington and Underwood, both American made. And just as the official files of the period abound with requests for automobile purchases and allowances, so are they replete with orders for typewriters or for their replacement and repair. The advent of the typewriter signaled a significant, if small-scale, revolution in governance. Along with electric lighting, electric fans, call bells, telephones, and duplicating machines, the manual typewriter transformed the government office in India in the decade or two before World War I, giving it a new technological élan while making it (not unlike the automobile) more remote from the public. The typewriter brought a new speed, efficiency, and orderliness to government business, just as electric fans replaced the creaking, hand-pulled *punkah*, and the telephone made contact between officials easier and speedier. As the threat to its authority grew, the colonial state, equipped with its own telephone switchboards, police radio vans, and motorcycle messengers, attained a new level of confidentiality but also a quickening response to emergency law-and-order situations.[17]

Typists and their writing machines were rapidly introduced into many areas of state activity—for instance, in local law courts, where they made possible accurate transcripts and multiple copies of court documents and proceedings. By 1914 most district courts in Madras Presidency had their own shorthand typists (as well as bicycle peons or messenger boys).[18] Typewriters also began to make a routine appearance in the revenue, educational, ecclesiastical, archaeological, police, and sanitation departments. Conversely, even a senior official, like the inspector general of prisons, might be castigated for presuming to hire a typist without prior sanction.[19] Shorthand reporting and typing became essential for the prosecution and conviction of political leaders and labor activists.[20] Like the automobile, by the 1920s no significant or status-conscious official would want to be without a typewriter and a typist at his command. Special committees, or periodic exercises like the decennial censuses, called for the appointment of additional typists and additional funds for typewriters.[21] Since typewriters were still relatively rare outside government, the state's possession of such a modern machine seemed to set it apart, to signal its superiority to the technologically backward society over which it presided.

The typewriter required a new range of office skills and a new personnel (female as well as male) accomplished in typing, taking shorthand, answering telephones, and operating switchboards. The old-fashioned scribe, who had served India's bureaucratic regimes for centuries, was increasingly obsolete. In a further extension of state business, examinations were held at regular intervals to test shorthand speeds and typing proficiency, with new pay scales and allowances introduced to reflect qualifications. The typing schools noted in chapter 3 drew much of their clientele from candidates for state employment. The typewriter became emblematic, too, of the continuity of government business, even in times of crisis. When an earthquake struck Bihar in 1934, it was reassuring (at least to one young British engineer) to see a government clerk still at work, in the street outside his wrecked office, "drumming on his typewriter in the full public gaze."[22] In August 1947 the government of India in New Delhi inherited a well-established bureaucracy in which the typewriter was an essential, even ubiquitous, tool. By contrast, the hastily convened and technologically impoverished government of Pakistan in its makeshift capital, Karachi, was desperately short of typewriters, as it was of most other items of office equipment.[23]

Subject to Technology

The colonial state was far from having a monopoly over the use of everyday technology, but there were ways in which it sought to use its privileged access to enhance its authority over its subjects, even to discipline and reform them. To illustrate this point, but also to highlight the limitations of such a presumption of power, the following section takes up two contrasting examples—state use of the bicycle and machines in prison.

As the bicycle came into widespread use in the West, it was adopted not just by the general public but also by state employees, including policemen, postmen, and health and sanitary workers. It was adopted in Britain and elsewhere in Europe by the armed forces: by the turn of the century the British Army alone had three thousand armed "cyclemen." Some of the troops who rode to war, if not into battle, on the western front during World War I did so on bicycles. It is not surprising that bicycles found similar service in India. There, too, they were considered a "useful accessory to modern warfare," as well as a means of keeping soldiers fit.[24] Cycles were supplied to units of both the British and Indian armies, and Indian troops equipped with bicycles fought in World War I, as at the Battle of the Somme. Bicycles also became part of the equipment of the Volunteers, racially exclusive, part-time regiments whose origins dated back to the anticolonial Rebellion of 1857–1858. Consisting of European and Anglo-Indian civilians, these units were mobilized during riots, strikes, and other disturbances to maintain order and assist the police and the regular army. By the late nineteenth century there were more than twenty Volunteer regiments in India; from about 1890 several, including those in Calcutta, Rangoon, and Bangalore, had "cyclist companies."[25] It does not appear, though, that these were ever used in conflict situations, and by the early 1920s, the cycle units were replaced by motorbikes and armored cars.

Bicycles enjoyed a similar vogue among India's police and paramilitary forces between the 1890s and 1920s. Putting policemen on bicycles promised to improve their mobility, speed the transmission of messages and commands, and facilitate armed intervention. In some respects the bicycle seemed poised to replace the more expensive horse and groom. City police stations were assigned bicycle orderlies, a duty for which "men mounted on bicycles" were deemed "more efficient and less costly." In some districts of Madras Presidency bicycles were introduced to enable policemen to move freely in rice-growing tracts where the quickest means of travel was along earth *bunds* (embankments), too narrow even for a horse.[26] But such changes did not go unchallenged. In 1923, when, on grounds of economy, the Madras government replaced the Rs 25 monthly horse allowance for sub-inspectors with a Rs 5 cycle allowance, the police inspector general protested—in vain—that this would be "a severe blow to the efficiency and well-being" of his rural subinspectors.[27]

However, the bicycle remained, in terms of state power, a vehicle of uncertain utility. It was believed essential in Madras city, where several riots occurred between 1918 and 1923, to retain a body of mounted constables for urban crowd control: no one was overawed by a man on a bicycle.[28] Part of a special paramilitary force, stationed in Malabar on the west coast of Madras Presidency to keep watch on the Mappillas and quash their periodic uprisings, was converted into a cycle corps in 1915. It was thought that bicycle messengers would allow rapid communication between the area's scattered rural settlements and that cyclists could transmit urgent news of incipient unrest. However, during the opening skirmishes of the ferocious Mappilla uprising of August 1921 a police messenger was pulled from his bicycle and killed. Another constable was beaten to death and his battered cycle carried off to the local mosque in triumph. In a clash with armed rebels a few days later, members of the cycle section fared badly, and, in the face of such vulnerability, the decision was taken to disband the unit. Thereafter, armored patrol cars and

FIGURE 6.1. Indian soldiers on bicycles during the Battle of the Somme on the western front in France in 1916. Image Q3983 by kind permission of the Imperial War Museum, London.

radios were used for surveillance and rapid response.[29] Bicycles thus became part of the small change of colonial governance. They had their uses—for routine police work or ferrying clerks to and from daily employment. But they proved of little value as a means of control and deterrence, and could even, as in Malabar in 1921, be a liability.

Prison technology reveals a somewhat similar ambivalence in the state's relationship with the machine. In the second half of the nineteenth century, where previously extramural labor had prevailed, prisoners were set to work machines within prison walls. An underlying argument for the creation of "jail industries" was financial (to generate income and cover prison costs), but the main rationale was that they fostered "habits of industry." Jails became innovative sites of mechanization, making carpets, mats, gunny bags, boots, shoes, and blankets. Some of the goods they produced, such as items of clothing, were for use within the prison, but many were supplied to other government departments or sold to the public. Learning new skills in jail, including the use of machines, was said to aid prisoners' rehabilitation and future employment. One commentator even argued that the prison was an essential site for disseminating new technology. "We are in the habit of complaining that the Hindus are impregnable to new processes," he wrote, "and hopelessly attached to the ancient systems of their country. Let us then introduce improvements, when the power of doing so is in our hands, by the instrumentality of convicts." In a society where technology remained primitive, the prison could become "the pioneer of progress."[30]

But this attempt to mold the lives of colonial penal subjects and speed the spread of machine technology ran into several obstacles. Rather than propagating new skills, the distribution of labor in prison tended to replicate the occupational hierarchy found outside it. So although tailoring, including the use of sewing machines, was adopted as a useful and productive form of labor in many jails, it tended to be work assigned to members of castes and communities who would anyway have engaged in tailoring outside.[31] Local manufacturers also protested that subsidized jail industries were undercutting their markets, eroding the profitability of their goods and violating state laissez-faire principles. In consequence, the use of steam-powered machinery in jails was banned in the 1880s and, though some power-driven industries survived, most prisons reverted to more elementary modes of production.[32] It was further argued on penal grounds that hard, repetitive labor was more punitive, and ultimately more deterrent, than the use of power-driven engines. Thus when Mary Frances Billington visited the Madras penitentiary in 1893 she found that mechanized rice milling was absent and, following conventional gender roles, women convicts were required to pound paddy for other convicts.[33] Indeed, as beriberi spread across monsoon Asia, many prison authorities deliberately encouraged hand pounding as being conducive to both health and discipline. Far from being "pioneers of progress" prisons became frequently places where arcane technologies were preserved.

In the 1890s, and especially following the disputed 1905 partition of Bengal, Indian jails began to receive political prisoners, many from educated, middle-class backgrounds. At the Port Blair penal settlement in the Andamans, where many convicted revolutionaries were sent, prison labor was given a new punitive purpose—to break the prisoners' political will and defiant attitude. V. D. Savarkar described how young men, "who in their lives had not done any physical labour," were made to work manually operated oil-crushing mills for several hours a day and were beaten if they failed to deliver their daily quota of oil. Even those, "tender in age and body," who were, or claimed to be, ill, earned no respite from this "racking toil." The oil mill (*kolhu*), associated like the rice-husking *dhenki* with peaceful and harmonious village life, became for the prisoners the mill that "ground down their lives."[34]

If at one level Indian nationalists of the *swadeshi* and Gandhian era sought to revalidate manual labor and restore preindustrial occupations, at another, in the contested confines of the colonial prison, precisely the same manually operated devices were annexed to a punitive penal regime, redolent of humiliation and hardship. Prisoners, and those outside the jail who spoke on their behalf, decried this use of human labor for tasks more readily and humanely performed by machines. Recounting his experiences as a political prisoner, one nationalist protested against "inhuman labour" in the jails, especially the use of the hand-driven *kolhu* to crush oilseeds. "In no other civilized country on the surface of the earth," he declared, "[are] such forms of labour . . . employed in jails."[35] In 1938, under a Congress ministry, the United Provinces jails reform committee recommended the abolition of human labor to operate oil mills and pound rice, tasks that should be done by machines or animals, not humans. Hand spinning alone was acceptable prison labor.[36]

Technology and the Expansionary State

Technology transformed the outward form of the state, quite as much as its inner ordering, and forced upon it new roles and responsibilities. This was a largely unwilling elaboration of colonial governance. For reasons of economy, from ideological adherence to laissez-faire, and from unwillingness to meddle more than necessary in the functioning of Indian society (if only for fear of provoking a popular backlash), the administration was disinclined to take on a new role of managing technology. Increasing the state's power in this way was even seen as conducive to greater corruption and oppression. And yet there were areas where a significant expansion of state power did occur, including those where it impinged upon the regulation and use of small machines.

Let us return to the streets, one of the main arenas were small machines were publicly put to use. From early in the twentieth century cyclists had to contend with a growing diversity and volume of traffic. Although the number of vehicles appears almost insignificant by today's standards, a modern sense of traffic, and the problems it created, had begun to emerge in India by 1914 and had become entrenched in state practice and public concern by the 1920s and 1930s. The most obvious factor in the growth of traffic, in India as elsewhere, was the arrival of new mechanized forms of transport—the tram, followed in rapid succession by the automobile, motorcycle, taxi, bus, and truck. As early as 1908 Bombay city had 276 automobiles on its roads and Madras 250. Calcutta had 202, rising by 1914 to 517 automobiles, 240 taxis, 152 motorcycles, and 21 trucks.[37] Small by European and North American standards, the volume of motor traffic in India increased exponentially after 1918 and, despite the Depression, surged during the 1930s. By 1936 British India had around 125,000 motor vehicles.[38] Despite petrol rationing and a dearth of civilian vehicles, traffic volumes continued to grow during World War II, reaching new heights after the war. In Madras the number of vehicles leapt from 18,653 in 1946 to 41,128 in 1948.[39]

But new technologies do not necessarily supersede older ones, and India's streets and highways were a scene of extraordinary and, to some observers, bewildering diversity.[40] Modern automobiles, trucks, and buses were in competition for road space with hand-pulled carts, ox wagons, horse carriages, pedestrians, cyclists, and stray animals. The apparent chaos of the street be-

FIGURE 6.2. A cycle-rickshaw driver waits for customers in the early morning traffic on the busy streets of Old Delhi. Author's photo.

came a metaphor for what many outsiders saw as the disorder of Indian society itself.[41] Although in evident decline, India's various horse-drawn passenger vehicles (victorias, *tongas, jutkas*, and so on), survived into the 1940s and beyond, in part because they catered to a less affluent clientele than automobiles and taxis.[42] There were some more obvious casualties. As late as the 1890s, there had been six hundred palanquins for hire in Calcutta. By 1919 the number had fallen to seventy and by 1928 to two.[43] But motor transport was not the sole beneficiary of the palanquin's demise. Originating in East Asia, hand-pulled rickshaws began to appear on the streets of Calcutta, Madras, and other Indian cities early in the twentieth century. As a cheap and convenient mode of urban transport, their numbers almost kept pace with the growing volume of automobiles and bicycles. In 1913 Madras had 482 cars to 4,413 bicycles and 2,900 rickshaws. By the mid-1930s there were 7,000 rickshaws on the city's streets, more than in Calcutta.[44] Pulling a rickshaw was among the most arduous and degrading of all urban occupations, but, despite public protests, the rickshaw survived to flourish during the fuel shortages of World War II, until increasingly superseded by the cycle rickshaw and auto rickshaw in the late 1940s.[45]

State involvement kept pace with the growth in traffic. At the level of municipal government, attempts were made to tax individual vehicles and their users—from dog carts and palanquins to cycles and motorbuses—according to scales determined for each class of vehicle. Many municipalities soon abandoned the bicycle tax as being too onerous and unpopular, but the Madras Corporation was one of the few to persist in levying a twice-yearly tax, issuing a small brass plate to each registered machine. But wholesale evasion followed and by the mid-1920s the number of licensed cycles fell far short of those actually in use. When stringent measures were introduced to enforce the tax, the number of registered bicycles rose dramatically—from under four thousand in 1927 to over twenty thousand in 1933. But even the Madras Corporation decided that the cost and effort of enforcing the bicycle tax was not worthwhile and concentrated its attention on motor vehicles instead.[46]

Legislation for the licensing of motor vehicles was introduced along metropolitan lines from the early years of the twentieth century. Although taxation was one incentive for this, no less compelling for the state was "the growing popularity of motor cars and the dangers attendant upon their use by reckless or unskilled persons."[47] However modest they may appear to us now, one spur to state involvement, and particularly police intervention, was the growing scale of traffic accidents and resulting fatalities. Across British India in 1933 there were 954 reported deaths due to road accidents and 6,611 serious injuries. By 1935 the figure had jumped to 1,309 and 9,621, respectively, and it continued rising in almost every province. Relative to the number of motor vehicles on India's roads (around 200,000), this was thought to be three times the accident rate in England and Wales and twice that in Germany. In the mid-1930s India was said to have the world's highest road accident rate, though the number of deaths and serious injuries was far below the United States, where 23,000 automobile-related deaths were recorded in 1924 alone.[48] Various explanations were given for India's high accident rate—the poor quality of road surfaces, careless driving and excessive speeding, the overloading of passenger and goods vehicles, the dilapidated state of overworked trucks and buses—but a lack of "road sense" was held to be a major factor.[49]

The blame for traffic accidents often fell on pedestrians and cyclists. Just as workers in rice mills were held responsible for the accidents that befell them, so were cyclists and pedestrians blamed for putting other road users at risk, their unmodern bodies and behavior imperiling the legitimate advance of motorized modernity. Since cyclists and pedestrians together were said to cause nearly half of all road accidents, "the necessity for instilling road sense" into them was, from the authorities' perspective, "a very urgent need."[50] From early in the twentieth century cyclists were among the principal victims of traffic accidents, as they ran into automobiles, taxis, and trams, and were fatally injured as a result of "reckless driving" by other road users or due to their own apparent folly. The newspapers of the period are full of sensational accounts (illustrated by photographs of wrecked automobiles and overturned trucks) reporting episodes like the "Serious Smash in Chowringhee" in June 1922, when two racing taxi drivers in Calcutta collided with a cyclist and the Tollygunge tram, or the "Flora Fountain Tragedy" in Bombay in December 1926 when a fifteen-year-old messenger boy was knocked down and killed by a speeding driver.[51] As if to deny their modest claims to modernity, bicycles and their riders were classed alongside ignorant pedestrians and lumbering oxcarts as among the hazards with which India's long-suffering motorists had to contend. In the discourse of modern traffic the cyclist was rated a lowly animal.

In order to combat crime, check on vehicles' roadworthiness, and maintain order on the streets, traffic became an increasing area of state responsibility. Police reporting of traffic accidents gives one indication of the growing scale of the perceived problem, as do attempts to prosecute offenders. As early as 1923 the Madras city police had a special staff to deal with traffic violations: in that year alone 1,637 cases were taken up and 783 individuals convicted. By 1940–41, across the province, there were 52,663 convictions under the Madras Traffic Rules, 21,892 under the provincial Motor Vehicles Act, and a further 163 under sections of the Indian Penal Code.[52] Rather like factory inspectors faced with recalcitrant rice-mill managers, police chiefs might lament the "dilatory and ineffective disposal of traffic prosecutions" and the inadequacy of the fines and sentences imposed; but the scale of the effort involved in attempting to regulate road traffic in India remains a remarkable (if barely noticed) aspect of late colonial policing.[53] At the same time, entrusting such extensive responsibility to the subordinate constabulary led to a proliferation of petty harassment and bribe taking. Cyclists, along with the drivers of automobiles, trucks, and buses, opted to pay a few rupees to an obstructive and threatening policeman rather than have their real or alleged misdemeanors punished through the courts. The growth of traffic helped fuel police rapacity and the coercive attributes of the everyday state.[54]

With respect to traffic, as with so many other areas of colonial governance, World War II ushered in an unprecedented crisis. Small machines—from typewriters to sewing machines—became scarce and much sought-after commodities. The shortage of petrol and the scarcity of motor vehicles for civilian use gave fresh impetus to the use of bicycles, even rickshaws. During the latter stages of the war and in its immediate aftermath, heavy military trucks, driven recklessly and at high speed, brought fresh carnage to Indian streets and pushed the accident rate in cities like Calcutta to unparalleled levels.[55] The principles of the free market, which had guided Britain's conduct of its economic business in India for more than a century, had weakened in the interwar years; now they crumbled. As the state sought to impose controls—even typewriter sales became subject to state regulation—so black marketeering and smuggling thrived. Acute food shortages, culminating in the "man-made" Bengal famine of 1943, which claimed an estimated three million lives, forced the central and provincial governments in India to introduce unparalleled restrictions on the sale, distribution, and pricing of essential food grains.

Up until this time state involvement in rice milling had been limited to the work of the factory inspectors and expressions of concern about the nutritional defects of white rice, and even there no action had been taken to curb the growing popularity of milled grain. However, with the loss in 1942 of Burma (a major source of India's rice) to the Japanese, with famine in Bengal in 1943, and with shortages throughout the country of essential commodities, including wheat and rice, governments were forced to act. Comprehensive controls were established, as in Madras, for the state procurement of rice, authorizing officials to buy grain directly from mill operators, who were now required to hold a special license. Further measures were imposed to restrict the amount of grain processed by each mill and (since heavy polishing was regarded as wasteful) the degree of milling permitted. The movement of grain and the market price of rice were also subject to state regulation.[56] In some areas rice, like fuel and cloth, was strictly rationed, and one effect of this was to force consumers to change their eating habits, as by switching from rice to wheat or maize. Nor were restrictions on the milling and sale of rice lifted when the war ended in 1945. In some respects state interventionism even increased. For instance, in 1950, an advisor to the central government's food ministry urged the nation's twenty thousand rice mills to switch to modern hulling machines, which produced less waste and gave a higher percentage of husked grain for the same quantity of paddy. India could thereby save an estimated forty thousand tons of rice a year.[57]

Once an emblem of small-scale and largely unregulated capitalism, rice mills had become a key area of state concern, subject to a much tighter regime of regulation than they, or many other factories, had ever previously experienced. This interventionism extended to the zealous punishment of mill managers and owners who chose to flout the new controls. Typical of the notices issued by the Bombay government, in December 1947, one was published for an individual fined Rs 200 for polishing rice without a license.[58] Apart from rampant smuggling and black marketeering, rice mills and rice depots became sites of conflict between consumers and the police sent to protect them. In one such incident in Madras in March 1946, police, goaded by stone throwing and the burning of two police bicycles, fired on protestors outside a wholesale rice store at Cheyyar in North Arcot: one man was killed and four others wounded.[59] Just as state policy after 1945 veered toward positive support for indigenous industries, like the manufacture of bicycles and sewing machines, so commonplace machines, like rice mills, entered a new era of state regulation. The everyday machine was fast becoming the state's machine.

The Political Road

An underlying theme of this discussion of everyday technology has been the question of resistance—an issue that has featured prominently in recent Indian historiography.[60] To what extent was there popular resistance to the coming of the modern machine? Or was the opposition that did occur more an effect of middle-class anticolonialism and technological nostalgia rather than a groundswell of subaltern dissent? Was hostility to the machine, rather than a species of Luddite atavism, merely a proxy for antipathies rooted in racial antipathy and class antagonism? The history of the machine, especially the machine outside the factory gates or away from the railroad tracks, has hitherto attracted so little comment in India that only preliminary observations are possible.

There is certainly some evidence to suggest an initial element of popular suspicion at the coming of the machine and modern technology during the nineteenth century. Smallpox vaccination—a pioneering medical technology—gave rise to extensive rumors about the evil nature of colonial intentions.[61] Likewise, it was suggested by some British commentators that the advent of the railroads and the locomotive caused such alarm that even the unrest of 1857 could be partly attributed to it—a claim that has also been strenuously denied.[62] Suspicion of modern technology, especially technologies associated with an alien and coercive state, reached a crescendo in the 1890s and early 1900s with the plague epidemic, when rumors abounded of the sinister purpose of hospitals, ambulances, and antiplague inoculations. Reports circulated of mysterious engines hidden in train sheds and devices to extract oil from Indian bodies. Europeans and Eurasians, vulnerable on their bicycles, were attacked merely on suspicion of being inoculators.[63] A concern, not confined to the subaltern classes, for the safety of the Indian body raged alongside popular fears about the malignancy of the machine.

But the crisis of the 1890s almost had the effect of finally clearing the air of lingering doubts. As machines became more widespread and common, India's subaltern classes seemed to show little overt opposition to the coming of such modest manifestations of technological modernity as sewing machines, bicycle, and rice mills—and one can see in this a precursor to recent acceptance of (and increasing demand for) such modern phenomena as x-rays, television, and cell phones. For many, especially among the artisan, laboring, and clerical classes, new technologies were acceptable, if not especially agreeable, extensions of traditional tools and older work regimes. Some technological innovations offered new possibilities of employment, an easier passage to work, perhaps, or the prospect of (marginally) higher wages and better working conditions. There was no *darzi* revolt against the Singer, no scribal insurrection against the Remington. That said, though, there was an often painful recognition that changing technology could bring radical changes to the relationship between peasants and urban workers on the one hand and landlords, industrial employers, and the state on the other. This was evident, for instance, on the streets of Indian cities, like Madras in December 1920, when striking mill workers found themselves confronted by trucks filled with armed police and strikebreakers. When stones were thrown and attempts made to stop the vehicles from entering the factory gates, the police opened fire, leaving a score of strikers dead or wounded.[64] A passage in Mulk Raj Anand's novel *Coolie* conveys the sense of powerlessness new technologies might create among those who did not command them. The mill hands in his story, having just learned that they are to be put on short time and their wages cut, turn toward the car of the factory owner. "They saw the long, black polished body of the Daimler swerve round. They rushed towards it, vaguely aware that the master of the mill was being driven away after pronouncing their doom. They would have fallen at his feet with joined hands if the car had not slid away."[65]

But the change was not all one sided. Even small machines could be deployed in ways that challenged, however temporarily, the power of the state and the ruling elite. When Gandhi took up the grievances of the indigo-growing peasants in Champaran District in Bihar in 1917, many of the advantages of modern technology appeared ranged against them. European planters demanded additional taxes from the hard-pressed peasantry, even to pay for their new automobiles. But when Gandhi began collecting evidence from the peasants, he found a surprising source of assistance in typists from the government survey department who volunteered their off-duty services to record the peasants' testimony.[66]

Bicycles became vehicles of opposition. Party activists cycled through city streets, handing out leaflets, and chanting nationalist slogans, or they pushed their bicycles in procession, portraits of leaders mounted on their handlebars. The bicycle stood for a new political connectedness. In Raja Rao's 1930s novel of rural south India, news of Gandhi's civil disobedience movement is ferried to the village of Kanthapura by waves of cyclists.[67] During provincial elections in 1937 bicycles were one of the means by which the Congress communicated its message to voters. The southern city of Madura mobilized a hundred cycle volunteers, the town of Erode another thirty.[68] During Gandhi's final fast in January 1948 more than five thousand cyclists advanced on Birla House, where he was staying, to hear a report on his declining health. As the twilight gathered they turned on their cycle lamps enabling Nehru to continue his impromptu speech, lighting up the garden as if with fireflies.[69] It is true, however, that bicycles had a less overtly insurrectionary role in India than elsewhere. They were never as strategically important there as they were in Vietnam, where they were essential to keeping Vietcong supply routes and information channels open. When Indian nationalists came closest to insurrection, as in the Quit India Movement

FIGURE 6.3. Sikh milkmen, with milk cans attached to their bicycles, participate in a *morcha* (protest) demonstration in Punjab in September 1946. Photo
by Margaret Bourke-White, reproduced by kind permission of Getty Images
(50875546).

of 1942, sabotage was primarily directed against the big, state-
supporting technologies of railroad tracks and telegraph wires,
and it was met in turn by armed force, including air strikes. Bicycles were of little use to either side.

But, whether through identification with the state, with European and Indian elites, or simply with the opposing party of

the day, the machine could constitute a convenient (as well as symbolic) target for attack. This was especially so on the streets where the modern machine was most conspicuously displayed or most obviously vulnerable. In India, too, despite the growing use of traffic police and paramilitary forces, the street remained a contested site, a place of uncertain state colonization. In moments of conflict the denizens of the street could change rapidly from being mere bystanders to being either victims or assailants. From the 1890s onward, in cities like Calcutta, Bombay, and Madras, trams, taxis, buses, and private automobiles (driven by Europeans or by well-to-do Indians) repeatedly came under attack during demonstrations and riots. Crowds, sometimes aided by elements of the so-called "criminal classes," threw stones, wrecked vehicles, and set them on fire.[70] Various factors were involved in these scenes of street violence—anger at the continued movement of traffic despite calls for protest demonstrations and strikes, hostility on racial or anitcolonial grounds to European officials, Eurasian tram drivers and ticket inspectors, or the affluence and indifference of Indian elites. Perhaps, as European businessmen and officials would have it, such violent eruptions were a result of the opportunist activities of "budmashes or ruffians with no ostensible means of subsistence, who are always prepared to create disturbance and commit crimes of violence."[71] But at other times incidents arose from a shared sense of outrage, as in Calcutta in March 1946, when furious onlookers attacked truck drivers and burned vehicles that had killed or injured innocent pedestrians and cyclists.[72]

By the 1930s and 1940s oppositional use of the street and the technology it contained had become relatively common, but it could still reach new levels of destructiveness. As British rule in India lurched toward the brutal denouement of partition, and state control over the street wavered, the flow or interruption of traffic took on a profoundly sinister character. The American photojournalist Margaret Bourke-White recalled the testimony of a Hindu tea-shop owner, Nanda Lal, to illustrate how Hindu-Muslim conflict erupted onto the streets of Calcutta on "Direct Action Day" on August 16, 1946. One of the first signs of impeding calamity was the sudden absence of traffic—with only a single, almost empty tram hurtling down Harrison Road. "It was the sight of that tram," Bourke-White wrote, "that confirmed Nanda Lal's fears that this day was to be unlike all other days." Normally it was "so crowded with commuters that they bulged from the platform and clung to the doorsteps and back of the car. Today there was hardly a passenger on board." Soon after, truckloads of Muslim *goondas* (thugs) arrived to attack Hindu shops (including Lal's), smashing rickshaws, setting automobiles on fire. Only gradually, after army tanks and machine guns had appeared on the streets, and an estimated five thousand deaths, was the return of normality signaled by the resumption of everyday traffic.[73]

Partition came barely a year after the "Great Calcutta Killing." It saw the reenactment of similar or still more violent scenes in which modern technologies of all kinds were implicated. In their haste to escape the carnage, tailors abandoned their sewing machines in the homes or on the verandahs where they had last been used. Trucks, automobiles, and motorbikes, as well as guns and bombs, were used as means to attack communal enemies. In some instances ropes were stretched across roadways to knock cyclists from their machines and kill them in cold blood. Overcrowded trains carrying refugees across the newly made border between India and Pakistan were attacked and their passengers massacred. Refugees in their thousands traveled by foot or huddled in oxcarts. Some carried their bicycles with them into exile, slung on the back of a bullock cart or truck; a privileged few made their escape by air.[74] In this maelstrom of violence the technology of the everyday continued to make its appearance, sometimes as an instrument of hatred, sometimes as a faint sign of hope.

×××××××××××××××

The relationship between the modern state and everyday technology was a complicated, multifaceted one, but no less important for that. The arrival of modern technologies—the technologies of the office, the army, the police, the courts, and even those of the factory and the street—could enhance the power of the state and those who served it. Modern technology could help make (but also unmake) modern subjects. Even such seemingly simple devices as sewing machines, bicycles, and typewriters could be pressed into state service. These could become emblems of authority or means by which the state sought to regulate and reform its subjects. The gradual extension of state power and responsibility that occurred after 1900, greatly accelerated by World War II, showed how even small machines like rice mills could become strategic sites in attempts to manage the economy and shape social practice.

Our concern here has been with small machines, not with those that were most powerful and persuasive, and yet one can conclude that India, like many colonies, did not submit easily to the authority—and the authoritarian capabilities—of the modern machine. It was less that Indians recoiled from the machine (though some undoubtedly did) than that the colonial state was in many respects too weak, its resources too slender, its ideological tools and material technologies too vulnerable, for effective command through the deployment of modern technology to be entirely possible. Technology was a two-edged weapon. The everyday technologies of the state, precisely because they were increasingly visible and accessible to others, could almost always be countered, subverted, or turned to oppositional ends. The proliferation of everyday technology and its regulation by the state might occasion corruption and resentment rather than compliance and control. Although the engagement of the state even with small machines—their making, distribution, licensing, and daily use—was undoubtedly growing as the colonial era ended, the state enjoyed no monopoly over how such technologies might be used or who might use them.

Epilogue

THE GOD OF SMALL THINGS

In 1997 the Indian writer and activist Arundhati Roy published a prize-winning novel entitled *The God of Small Things*.[1] Her book is not about technology, or even the history of technology, though some everyday technologies (a car, the cinema) do appear in it. But it is about India and the ways in which small objects and seemingly small events and emotions shape people's lives and cumulatively have a profound effect on their existence. The implication is not that small things are necessarily divine, though in India, as elsewhere, that association can exist, even with respect to machines. It is rather that small things can be "godly" in the sense of being of primary, even paramount, importance when it comes to daily lives and the everyday objects with which people surround themselves. I take the idea of the small as "godly" as suggestive for a reading of small-scale technologies in their global, imperial, and, not least, Indian contexts. I see this focus on small, everyday technology as a much-needed corrective to the excessive emphasis on large-scale technological systems that has dominated the discussion of technology in relation to India and the non-Western world—railroads, telegraphs, irrigation networks, and big dams. It is a means by which to highlight instead the material importance and social significance of "small things."

The question is not whether small technologies were "godly" in a religious or ritual sense, or even "beautiful" in the way that E. F. Schumacher influentially deployed that term.[2] Nor were they even necessarily "appropriate" in the sense in which many advocates of "appropriate technology" have used that adjective. Gandhi was not alone in believing that some small machines, like rice mills, were highly inappropriate for a poor, underemployed country like India, or, like bicycles, catered to a needless craving after speed. The point, rather, is that such small machines impacted on people's lives in highly significant ways, and acquired, by their use, and even by their denial, a social meaning and a cultural traction that belied their seemingly insignificant nature. The mundane could be momentous. This is not to say that small machines of the kind discussed here were *always* agents of material progress and social transformation. In some contexts they certainly were, but in others they served on the contrary to reinforce, to inscribe anew, the presumptions and privileges of race, class, and gender.

In the global history of small objects locality matters greatly. By virtue of their accessibility, affordability, and mobility, small technologies became closely entangled with the occupations, values, aspirations, even the emotions of the local populations with which they became embedded. The history of modern technology is more than a history of globalization, as conventionally understood, or a history of the rise to global ascendency of a technologically empowered West. Commodities, which by the late nineteenth and early twentieth centuries were becoming global goods and entering everyday use around the world, were simultaneously becoming local goods, subject to local needs and desires. Purveyors of local and not just universal meanings, they could morph into instruments of empowerment and subversion. They became local emblems, the subjects of individual and collective imaginaries as much as work tools or mere items of consumption.

This might be true everywhere, but in India everyday technology occupied—and still continues to occupy—a pivotal position. Some of the reasons for this have already been alluded to. India's chronic poverty and its apparent technological backwardness, the extreme nature of its social differentiation, the constraints and ambitions of colonial rule, the rise of Indian capitalism and the pursuit of *swadeshi* ideals, the deep-seated uncertainty about the modern machine that manifested itself in harkening back to a preindustrial age—all combined to afford the small machine an exceptional role and a strategic significance. In colonial rhetoric and nationalist polemic, in state policy and in the materiality of production, the small machine played a symbolic, as well as practical, role. India's quest for modernity was as much about an engagement with small things as an entanglement with large ones.

That duality was perhaps not so immediately evident with the end of World War II in 1945 and Indian independence two years later. India seemed to emerge from the war with a new confidence in its technological modernity. An advertisement for India's National Savings in early 1946 was indicative of this, declaring:

> The heavy rumble and clatter of caterpillar tracks echo across the land but not to the roar of deadly guns. . . . Across India's wide acres we see a great awakening. Gone are the days of the wooden plough, the spade and the sickle with which our cultivators obtained from the good earth too little with too great a labour. In their place great steel ploughs, harvesters, binders and tractors make their toil easier and force richer yield[s] from the soil. . . . This is a picture of tomorrow as seen today.[3]

In Jawaharlal Nehru India acquired a prime minister who, unlike Gandhi, his mentor and nemesis, believed emphatically in the benefits that modern science and technology could confer upon India. "I . . . have worshipped at the shrine of science and counted myself one of its votaries," he told India's National Academy of Sciences in March 1938. "Who . . . can afford to ignore science today? At every turn we have to seek its aid and the whole fabric of the world today is of its making."[4] Even by the time that Nehru became chair of the National Planning Committee, set up by the Congress in 1938, the technological imagination of many of India's intellectuals and entrepreneurs had leapfrogged the problem of how to produce bicycles and sewing machines and was fixated instead on making automobiles and airplanes. With independence attained, and his ardor strengthened rather than diminished by the ordeal of partition, Nehru focused his attention—and that of his government—on five-year plans, industrialization, scientific development, and the urgency of rapid technological progress.[5] Despite occasional claims to the contrary, in Nehru's grand vision it was "big technology" and "big machines" that mattered—steel mills, hydroelectric dams, nuclear reactors—not the trivia of small-scale technology. Nehru spoke eloquently in 1954 of big dams as the "temples of the new age" and observed that "small minds" and "small nations" could not undertake "big works." "When we see big works," he declared, "our stature grows with them, and our minds open out a little."[6] As noted earlier, to Nehru it was a matter of regret that by 1960 India had only just fully entered the bicycle age, not yet that of nuclear power and jet planes.

And yet it could be argued that it was the continuing progress of small technology (far more than the Gandhian option of the *charka* or a return to hand-husked rice) that helped bring slow but significant change to India. As any journey around an Indian city or through the Indian countryside will demonstrate, bicycles remain an essential mode of personal transport and an invaluable means of moving all manner of small goods. Even in an era of economic growth and middle-class affluence, of motorbikes, motorscooters, and automobiles, the bicycle and its offshoots, such as the cycle rickshaw and the pedal-driven cart, remain essential. The spread of bicycles, and their production on a scale unmatched by colonial-era imports (with more than seven million bicycles a year made in India in the 1990s), contributed, however unspectacularly, to the growing physical and social mobility of rural populations, of women and the ex-untouchables, easing access to education, increasing or diversifying employment opportunities.[7] Typewriters may have largely (but not yet entirely) disappeared from Indian offices and sidewalk encampments, but manual or electric sewing machines still whirl away in roadside stalls and tailors' shops. Rice mills (and their analogues like flour mills) have continued to proliferate—their hum to be heard in many an urban backstreet and village lane—and have conceded no ground to attempts made since the 1930s to revive and popularize hand-husked rice. Villages, towns, and cities across India testify to the utility, the versatility, the ubiquity of the small machine.

While many of independent India's large industries and grand engineering projects have stalled, failed to deliver their grand Nehruvian promise, or, like big dams, become burdened with high environmental and social costs, small machines, like the power-driven water pumps that have been responsible for the greening of so much of the Indian countryside, have greatly impacted on livelihoods and landscapes. Many recent technology-related developments can be seen prefigured in an earlier history of everyday technology. The rise of the Singer sewing machine anticipated and facilitated the phenomenal growth of South Asia's massive garment industry and the accompanying sweatshops, many of them reliant, as in present-day Bangladesh, on underpaid female labor. There is a tenuous but not entirely fatuous connection between the small numbers of Eurasian women typists employed in government offices and business houses in the early twentieth century and the large numbers of women who now work in India's international call centers. Even the problem of traffic, first made evident a century ago on what may appear to us now as a curiously miniature scale, has lost none of its significance given the soaring accident rate and mortality on India's now-congested streets and crowded highways. And just as the use of bicycles and sewing machines became fairly rapidly established in city, small-town, and village society, demonstrating that presumptions of Indian technological ineptitude or inaptitude were wholly misguided, so in recent decades have cheap and convenient cell phones proliferated across India, claiming more than three hundred million users by 2006.[8] Small machines may not have solved India's underlying problems, endemic poverty and rapid population growth among them, but the history of the last sixty years has shown the importance of making, selling, and imagining small things as well as large ones. The god of small things is not yet dead.

The point is not that these technological developments are necessarily right and desirable, or "appropriate" in the conventional sense. Many manifestations of what have here been called

FIGURE E.1. A roadside tailor at work in Old Delhi in 2008. Author's photo.

everyday technology have not been conducive to wealth or well-being, as critics of technological modernity in India were quick to anticipate decades ago. Many, indeed, have proved instruments of exploitation and have helped to keep the poor needy, hungry, and oppressed. But they are cogent demonstrations of India's complicated—often muted, quizzical, antagonistic—engagement with technological modernity as well as its distinctive capacity to be inventive and adaptive, to absorb new technologies (without saying farewell to the old), and to put them to its own practical and ideological uses. Even demonstrably foreign machines could become indigenized and made integral to ongoing social and economic change. The history of everyday technology in India can be rendered in many ways—as a history of colonialism and development, of postcolonialism and globalization—but it is also a history of India's struggle with the technological opportunities and dilemmas that lie at the heart of all our modernities.

Acknowledgments

I have incurred many debts in researching and writing this book. My first and foremost debt of gratitude is to the Economic and Social Research Council in the United Kingdom, which funded a three-year research professorship for a project on "Everyday Technology in Monsoon Asia, 1880–1960." This made possible a substantial part of the research for this book and funding for a workshop on "Everyday Technology" at the University of Warwick in March 2010. My thanks to the workshop participants for the stimulus their work and comments provided. I am indebted to the University of Chicago for a visiting professorship in 2008 that enabled me to use the invaluable resources of the Regenstein Library and visit the Wisconsin Historical Society at Madison for its collection of Singer papers. Research on bicycles and typewriters, respectively, was made possible through the Raleigh Archives at the Nottinghamshire County Records Office and the Godrej Archives in Mumbai (with special thanks to Vrunda Pathare for her help). I have profited greatly, too, from the India Office collections at the British Library (including the Newspaper Library at Colindale), from the Library of the School of Oriental and African Studies in London, and from the University of Warwick Library: sincere thanks to the staff of all three.

Earlier versions of the material presented here have appeared in print as "Global Goods and Local Usages: The Small World of the Indian Sewing Machine, 1875–1952," *Journal of Global History* 6 (2011): 407–29; with Erich DeWald, "Cycles of Empowerment," *Comparative Studies in Society and History* 53 (2011): 971–96; "Technology and the New Subalternity," *Cadernos de Estudos Culurais* 3 (2011): 27–36; and "The Problem of Traffic: The Street-Life of Modernity in Late-Colonial India," in "Everyday Technology in

South and Southeast Asia," edited with DeWald, special issue, *Modern Asian Studies* 46 (2012): 119–42. That issue of *MAS* included an introduction in which Erich DeWald and I sketched out some of the issues explored at greater length here.

Work on this project has been helped by advice and suggestions from a large number of scholars, notably Erich DeWald, who collaborated with me on the "Everyday Technology" project, and Indira Chowdhury, who organized archival research assistance in India. I also would like to thank my Warwick colleagues Maxine Berg, Margot Finn, Anne Gerritsen, David Hardiman, Sarah Hodges, Claudia Stein, Tracy Horton (for her administrative expertise), and, now at Leicester, Clare Anderson, an unfailing source of pictures and references. Thanks, too, to Itty Abraham, Michael Adas, Shahid Amin, Dipesh Chakrabarty, Vinayak Chaturvedi, Greg Clancey, Markus Daechsel, Faisal Devji, Prasenjit Duara, David Edgerton, Prashant Kidambi, John Krige, Karen Leonard, Suzanne Moon, Chris Pinney, Satadru Sen, Chandak Sengoopta, Sanjay Subrahmanyam, Sarah Teasley, and Sylvia Vatuk, to the anonymous reviewers for Chicago, and to V. R. Muraleedharan for arranging a visit to a rice mill in Thanjavur. I am grateful to Julie Snook for drawing the map. Special thanks are due to Adrian Johns as series editor and Karen Merikangas Darling at the University of Chicago Press, who have helped to mold this book into its final form. Finally, since the history of everyday technology has proved singularly autobiographical, I want to thank my parents, May and John Arnold, who taught me to type, sew, and ride a bike, and express my heartfelt thanks to Juliet Miller for her fortitude and good humor. Without her love and support this book would not have been possible—nor have seemed half so worthwhile.

Notes

Introduction

1. Wiebe E. Bijker and John Law, "General Introduction," in *Shaping Technology / Building Societies: Studies in Sociotechnical Change*, ed. Wiebe E. Bijker and John Law (Cambridge, MA: MIT Press, 1992), 3.

2. Partha Chatterjee, *The Nation and Its Fragments: Colonial and Postcolonial Histories* (Princeton, NJ: Princeton University Press, 1993), 5.

3. Bijker and Law, "General Introduction," 3.

4. "TPW," "The Cycling Experience," *Scottish Review* 29 (1897): 57.

5. Frank Dikötter, *Things Modern: Material Culture and Everyday Life in China* (London: Hurst, 2007), 8.

6. David Edgerton, "Creole Technologies and Global Histories: Rethinking How Things Travel in Space and Time," *Journal of History of Science and Technology* 1 (2007): 75–112.

7. The significance of this ritual and the Vishwakarma festival has been much debated, whether as indicating the persistence of custom and the "godly" qualities of workmen's tools or as a site of resistance to workplace discipline. Leela Fernandes, *Producing Workers: The Politics of Gender, Class, and Culture in the Calcutta Jute Mills* (Philadelphia: University of Pennsylvania Press, 1997), 102–4; and Dipesh Chakrabarty, *Provincializing Europe: Postcolonial Thought and Historical Difference* (Princeton, NJ: Princeton University Press, 2000), 76–78. My point is simply that such rituals facilitated the incorporation and normalization of modern machines.

8. *Hindu* (Madras), January 16, 1904, 8.

9. Ravinder Kaur, *Performative Politics and the Cultures of Hinduism: Public Uses of Religion in Western India* (Delhi: Permanent Black, 2003), 76–77.

10. As in the deliberations of the Indian planning committee: K.

T. Shah, *National Planning, Principles and Administration* (Bombay: Vora and Co., 1948).

11. N. Gangulee, *Health and Nutrition in India* (London: Faber and Faber, 1939), 146; [M. N. Saha], editorial, *Science and Culture* 1 (1935): 3.

12. Rabindranath Tagore, *My Reminiscences* (London: Macmillan, 1991), 144; M. K. Gandhi, *The Bhagavad Gita* (Delhi: Orient Paperbacks, 1988), 133.

13. For a state-centred interpretation of India's technological modernity, see Gyan Prakash, *Another Reason: Science and the Imagination of Modern India* (Princeton, NJ: Princeton University Press, 1999), 160–70.

14. Glyn Barlow, *Industrial India* (Madras: G. A. Natesan, 1907); Alfred Chatterton, *Agricultural and Industrial Problems in India* (Madras: G. A. Natesan, 1903).

15. *M. K. Gandhi: "Hind Swaraj" and Other Writings*, ed. Anthony J. Parel (Cambridge: Cambridge University Press, 1997), chaps. 6 and 12.

16. E.g., *Bombay Chronicle*, June 1, 1938, 12.

Chapter One

1. Benedict Anderson, *Imagined Communities: Reflections on the Origin and Spread of Nationalism* (London: Verso, 1983); Ronald Inden, *Imagining India* (Oxford: Blackwell, 1990); Gyan Prakash, *Another Reason: Science and the Imagination of Modern India* (Princeton, NJ: Princeton University Press, 1999), 201.

2. Michael Adas, *Machines as the Measure of Men: Science, Technology, and Ideologies of Western Dominance* (Ithaca, NY: Cornell University Press, 1989), 204.

3. Rokeya Sakhawat Hossain, *"Sultana's Dream" and "Padmarag": Two Feminist Utopias*, ed. Barnita Bagchi (New Delhi: Penguin, 2005), 8, 12–13.

4. Bagchi, introduction to ibid., ix, xiii.

5. Ibid., 8.

6. Markus Daechsel, *The Politics of Self-Expression: The Urdu Middle-Class Milieu in Mid-Twentieth Century India and Pakistan* (London: Routledge, 2006), 134.

7. Nirad C. Chaudhuri, *The Autobiography of an Unknown Indian* (New York: Macmillan, 1951), 83, 176.

8. *M. K. Gandhi: "Hind Swaraj" and Other Writings*, ed. Anthony J. Parel (Cambridge: Cambridge University Press, 1997), 36, 107, 111.

9. *Collected Works of Mahatma Gandhi* [*CWMG*] (New Delhi: Publications Division, Ministry of Information and Broadcasting, 1966), 25:251.

10. Editorial, *Science and Culture* 1 (1935): 3–4.

11. Shiv Visvanathan, *Organizing for Science: The Making of an Indian Research Laboratory* (Delhi: Oxford University Press, 1985), 97, 105. Saha's vision is outlined in Prakash, *Another Reason*, 191–98.

12. Richard Drayton, *Nature's Government: Science, Imperial Britain, and the "Improvement" of the World* (New Haven, CT: Yale University Press, 2000).

13. Edward Thornton, *India: Its State and Prospects* (London: Parbury, Allen, 1835), 65–69.

14. Harriet Martineau, *Suggestions towards the Future Government of India*, 2d ed. (London: Smith, Elder, 1858), 32.

15. Richard Temple, *India in 1880*, 2d ed. (London: John Murray, 1881), 100.

16. Bernard S. Cohn, *Colonialism and Its Forms of Knowledge: The British in India* (Princeton, NJ: Princeton University Press, 1996), chap. 1.

17. "Madras Exhibition," *Calcutta Review* 26 (1856): 265–84.

18. Ibid., 266.

19. Madras Revenue Proceedings, Government Order 29 (hereafter cited as GO), June 16, 1868, India Office Records, British Library, London (hereafter cited as IOR).

20. For a critical evaluation of India's industrial exhibitions, see Glyn Barlow, *Industrial India* (Madras: G. A. Natesan, 1907), chap. 3.

21. Manu Goswami, *Producing India: From Colonial Economy to National Space* (Chicago, IL: University of Chicago Press, 2004), 47, 103.

22. David Nye, *American Technological Sublime* (Cambridge, MA: MIT Press, 1994).

23. G. W. MacGeorge, *Ways and Works in India* (Westminster, UK: Archibald Constable, 1894), 1.

24. Ibid., 1–2.

25. George C. M. Birdwood, *Paris Universal Exhibition of 1878: Handbook to the British Indian Section* (London: Offices of the Royal Commission, 1878), 49, 51.

26. George C. M. Birdwood, *Sva* (London: Lee Warner, 1915), 62, 64.

27. Ibid., 75.

28. Leo Marx, *The Machine in the Garden: Technology and the Pastoral Ideal in America* (New York: Oxford University Press, 1964).

29. R. V. Russell, *The Tribes and Castes of the Central Provinces of India* (London: Macmillan, 1916), 2:470.

30. Edgar Thurston, *Castes and Tribes of Southern India* (Madras: Government Press, 1909), 3:9.

31. Thurston, *Castes*, 5:102.

32. *Southern India: Its History, People, Commerce, and Industrial Resources*, comp. Somerset Playne (London: Foreign and Colonial Compiling and Publishing Co., 1915).

33. Thurston, *Castes*, 3:119–20; ibid., 4:294.

34. W. Crooke, *Natives of Northern India* (London: Archibald Constable, 1907), 147, 149.

35. Tirthankar Roy, "Foreign Trade and the Artisans in Colonial India: A Study of Leather," *Indian Economic and Social History Review* 31 (1994): 461–90; Roy, "Home Market and the Artisans in Colonial India: A Study of Brass-Ware," *Modern Asian Studies* 30 (1996): 357–85; and Roy, *Traditional Industry in the Economy of Colonial India* (Cambridge: Cambridge University Press, 1999).

36. Alfred Chatterton, *Industrial Evolution in India* (Madras: "Hindu," 1912), 22.

37. Charles Doyley, *The European in India: From a Collection of Drawings* (London: Orme, 1813), plate 13; H. A. Rose, *A Glossary of the Tribes and Castes of the Punjab and North-West Frontier Province* (Lahore: "Civil and Military Gazette" Press, 1911), 2:223.

38. Charles Allen, ed., *Plain Tales from the Raj: Images of British India in the Twentieth Century* (London: André Deutsch, 1975), 107.

39. There is an extensive literature on the spread of sewing machines in non-Western societies: for an example with close parallels to India, see, Uri M. Kupferschmidt, "The Social History of the Sewing Machine in the Middle East," *Die Welt des Islams* 44 (2004): 195–213.

40. J. Forbes Watson, *The Textile Manufactures and Costumes of the People of India* (London: William H. Allen, 1867), 4, 12.

41. Robert L. Hardgrave, *The Nadars of Tamilnad: The Political Culture of a Community in Change* (Berkeley, CA: University of California Press, 1969), 59–70; Himani Bannerji, "Textile Prison: The Discourse

of Shame (*lajja*) in the Attire of the Gentlewoman (*bhadramahila*) in Colonial Bengal," *South Asia Research* 13 (1993): 27–45.

Chapter Two

1. Arnold J. Bauer, *Goods, Power, History: Latin America's Material Culture* (Cambridge: Cambridge University Press, 2001), chap. 5.

2. "General Introduction," in *The Social Construction of Technological Systems: New Directions in the Sociology and History of Technology*, ed. Wiebe E. Bijker, Thomas P. Hughes, and Trevor J. Pinch (Cambridge, MA: MIT Press, 1987), 3.

3. Andrew Godley, "Global Diffusion of the Sewing Machine, 1850–1914," *Research in Economic History* 20 (2001): 22–23.

4. Andrew Gordon, "Selling the American Way: The Singer Sales System in Japan, 1900–1938," *Business History Review* 82 (2008): 694.

5. Robert Bruce Davies, *Peacefully Working to Conquer the World: Singer Sewing Machines in Foreign Markets, 1854–1920* (New York: Arno Press, 1976), v.

6. *Report of the Indian Tariff Board on the Sewing Machine Industry* (Bombay: Government Central Press, 1947), 12.

7. A. Vaidyanathan, "The Indian Economy since Independence (1947–70)," *The Cambridge Economic History of India*, ed. Dharma Kumar and Meghnad Desai, vol. 2, *c. 1757–c. 1970* (Cambridge: Cambridge University Press, 1983), 965. Bicycles rose in the same period from 750 to 3,000 per million (ibid.).

8. J. Masselos, "The Discourse from the Other Side: Perceptions of Science and Technology in Western India in the Nineteenth Century," in *Writers, Editors and Reformers in Social and Political Transformations of Maharashtra, 1830–1930*, ed. N. K. Wagle (New Delhi: Manohar, 1999), 122.

9. N. K. Adyanthaya, *Report of an Enquiry into the Family Budgets of Industrial Workers in Madras City* (Madras: Superintendent, Government Press, 1938), 44–47.

10. Alfred Chatterton, *Industrial Evolution in India* (Madras: "Hindu," 1912), 21–22.

11. Basil Mathews, *India Reveals Itself* (London: Oxford University Press, 1937), 178.

12. Rudyard Kipling, *Kim* (London: Macmillan, 1981), 214.

13. Mulk Raj Anand, *Untouchable* (New Delhi: Penguin, 2001), 35.

14. *The Webbs in Asia: The 1911–12 Travel Diary*, ed. George Feaver (Basingstoke, UK: Macmillan, 1992), 217, 289.

15. *Census of India, 1911*, vol. 14, *Punjab*, part 1, *Report* (Lahore: "Civil and Military Gazette" Press, 1912), 507.

16. Gareth Stedman Jones, *Outcast London: A Study in the Relationship between Classes in Victorian Society* (Harmondsworth, UK: Penguin, 1976), 22, 107–11.

17. *Census of India, 1911*, vol. 14, *Punjab*, part 1, *Report*, 507.

18. Margaret Read, *From Field to Factory: An Introductory Study of the Indian Peasant Turned Factory Hand* (London: Student Christian Movement, 1927), 22.

19. Athelstane Baines, *Ethnography (Castes and Tribes)* (Strasbourg, Ger.: Trübner, 1912), 98.

20. Expenditure ranged from Rs 60 to Rs 80 for a new bicycle to running costs of up to Rs 2 a month. Office of the Economic Adviser to the Government of India (Ministry of Commerce), *Report of an Enquiry into the Family Budgets of Middle Class Employees of the Central Government* (Delhi: Manager, Government of India Publications Branch, 1949), 100, 133, 297.

21. *Illustrated Weekly of India*, April 19, 1939, 73.

22. In 1934, 26,759 bicycles were registered in Madras, a city (in 1931) of 647,230 people, equal to one bicycle for every 24 people. The actual number of cycle users is likely to have been higher. *Administration Report of the Corporation of Madras, 1934–35* (Madras: Superintendent, Government Press, 1935), 16.

23. *Jawaharlal Nehru on Science and Society: A Collection of His Writings and Speeches*, ed. Baldev Singh (New Delhi: Nehru Memorial Museum and Library, 1988), 221.

24. *Report on the Administration of the Police of the Madras Presidency, 1938* (Madras: Superintendent, Government Press, 1939), 58; *Annual Report on the Police of the City of Bombay, 1939* (Bombay: Government Central Press, 1940), 3.

25. *Bombay Chronicle*, December 24, 1926, 11.

26. *Indian Industrial Commission: Minutes of Evidence, 1916–17*, vol. 1, *Delhi, United Provinces and Bihar and Orissa* (Calcutta: Superintendent of Government Printing, India, 1917), 20.

27. Remington Rand to Government of India, Home Department,

November 4, 1939, Home (Public) 45/30/39, National Archives of India, New Delhi (hereafter cited as NAI).

28. *Madras Mail*, January 8, 1903, 9.

29. See esp. Friedrich A. Kittler, *Gramophone, Film, Typewriter*, trans. Geoffrey Winthrop-Young and Michael Wutz (Stanford, CA: Stanford University Press, 1999); Rubén Gallo, *Mexican Modernity: The Avant-Garde and the Technological Revolution* (Cambridge, MA: MIT Press, 2005), chap. 2.

30. Frank Dikötter, *Things Modern: Material Culture and Everyday Life in China* (London: Hurst, 2007), 126.

31. Hilton Brown, *Parry's of Madras: A Story of British Enterprise in India* (Madras: Parry, 1954), 161; H. E. A. Cotton, *Calcutta: Old and New: A Historical and Descriptive Handbook to the City* (Calcutta: Newman, 1907), iv.

32. Rohinton Mistry, *A Fine Balance* (London: Faber and Faber, 1996), 558.

33. *Illustrated Weekly of India*, July 26, 1936, 13.

34. Bruce Bliven, *The Wonderful Writing Machine* (New York: Random House, 1954), 25.

35. List of locations given on letter heading from Remington Rand to Government of India, Home Department, November 4, 1939, Home (Public) 45/30/39, NAI.

36. David Ludden, "Craft Production in an Agrarian Economy," in *Making Things in South Asia: The Role of Artist and Craftsman*, ed. Michael W. Meister (Philadelphia, PA: University of Pennsylvania, 1988), 103–13.

37. A detailed account of the milling process is given in Edward B. Vedder, *Beriberi* (London: Bale, Sons and Danielsson, 1913), chap. 5.

38. *Census of India, 1911*, vol. 14, *Punjab*, part 1, *Report*, 506.

39. L. S. S. O'Malley, *Modern India and the West: A Study of the Interaction of Their Civilizations* (London: Oxford University Press, 1941), 255.

40. *Annual Report on the Working of the Indian Factories Act in the Punjab, 1924* (Lahore: Superintendent, Government Printing, 1925), vii, xi.

41. Alfred Chatterton, "Industrial Occupations," *Census of India, 1911*, vol. 12, *Madras*, part 1, *Report* (Calcutta: Superintendent of Government Printing, 1913), 199.

42. On Morley's intervention, see *Indian Industrial Commission, 1916–18: Report* (Calcutta: Superintendent of Government Printing, India, 1918), chap. 8.

Chapter Three

1. For some indication of how race and gender together might inform technology, see Judith A. Carney, *Black Rice: The African Origins of Rice Cultivation in the Americas* (Cambridge: Harvard University Press, 2001).

2. Tony Ballantyne and Antoinette Burton, "Introduction: Bodies, Empires, and World Histories," in *Bodies in Contact: Rethinking Colonial Encounters in World History*, ed. Tony Ballantyne and Antoinette Burton (Durham, NC: Duke University Press, 2005), 6.

3. Partha Chatterjee, *The Nation and Its Fragments: Colonial and Postcolonial Histories* (Princeton, NJ: Princeton University Press, 1993), 20.

4. Karin Hausen, "Technical Progress and Women's Labour in the Nineteenth Century: The Social History of the Sewing Machine," in *The Social History of Politics: Critical Perspectives in West German Historical Writing since 1945*, ed. Georg Iggers (Leamington Spa, UK: Berg, 1985), 259–81.

5. Tim Putnam, "The Sewing Machine Comes Home," in *The Culture of Sewing: Gender, Consumption and Home Dressmaking*, ed. Barbara Burman (Oxford: Berg, 1999), 269–83.

6. On Singer marketing practices, see Andrew Godley, "Selling the Sewing Machine around the World: Singer's International Marketing Strategies, 1850–1920," *Enterprise and Society* 7 (2006): 266–313.

7. John Mitchell to Singer, London, April 20, 1888, box 89, folder 3, Singer Archive, Wisconsin Historical Society, Madison (hereafter cited as SA).

8. D. Davidson to George R. McKenzie, May 20, 1884, box 88, folder 8, SA.

9. Smithsonian Institution, Washington, DC, SIL10-679-001. On Singer's "civilizing mission," see Robert Bruce Davies, *Peacefully Working to Conquer the World: Singer Sewing Machines in Foreign Markets, 1854–1920* (New York: Arno Press, 1976).

10. Jesse S. Palsetia, *The Parsis of India: Preservation of Identity in Bombay City* (Leiden, Neth.: Brill, 2001).

11. N. M. Patell to McKenzie, November 17, 1881, box 88, folder 8, SA; Patell to McKenzie, January 11, 1884, ibid.

12. Patell to McKenzie, July 3, 1883, ibid.

13. Mitchell to Singer, London, April 20, 1888, box 89, folder 3, SA.

14. On the perceived difficulties of the Indian market for American manufacturers and salesmen, see Henry D. Baker, *British India with Notes on Ceylon, Afghanistan, and Tibet* (Department of Commerce, Special Consular Reports, no. 70) (Washington, DC: Government Printing Office, 1915). Other Asian markets could also be problematic. See Andrew Gordon, "Selling the American Way: The Singer Sales System in Japan, 1900–1938," *Business History Review* 82 (2008): 671–99.

15. Patell to Singer, July 12, 1886, box 88, folder 8, SA; "List of Offices in India, Burma and Ceylon," 1905, box 89, folder 7, SA.

16. Edward Lang to Singer, London, July 17, 1883, box 88, folder 8, SA; Lang to McKenzie, August 10, 1883, ibid.; Davidson to McKenzie, December 29, 1883, ibid.

17. Fred. B. Fisher, *India's Silent Revolution* (New York: Macmillan, 1919), 45.

18. Patell to McKenzie, November 20, 1883, box 88, folder 8, SA.

19. Davidson to McKenzie, December 29, 1883, box 88, folder 8, SA; Patell to McKenzie, January 16, 1885, ibid.

20. Patell to McKenzie, January 16, 1885, ibid.

21. Smithsonian Institution, SIL 10-679-001.

22. Uri M. Kupferschmidt, "The Social History of the Sewing Machine in the Middle East," *Die Welt des Islams* 44 (2004): 204–9.

23. See the photograph of the Parsi Ladies' School in Palsetia, *Parsis*, 368.

24. Kupferschmidt, "Social History," 213; Mona L. Russell, *Creating the New Egyptian Woman: Consumerism, Education, and National Identity, 1863–1922* (Basingstoke, UK: Palgrave Macmillan, 2004), chaps. 6 and 8.

25. Anshu Malhotra, *Gender, Caste, and Religious Identities: Restructuring Class in Colonial Punjab* (New Delhi: Oxford University Press, 2002), 137.

26. *A Glossary of the Tribes and Castes of the Punjab and North-West Frontier Province*, comp. H. A. Rose (Lahore: "Civil and Military Gazette" Press, 1911), 2:378–79; S. T. Hollins, *The Criminal Tribes of the United Provinces* (Allahabad: Superintendent, Government Press, United Provinces, 1914), chap. 24.

27. Bejoy Shankar Haikerwal, *Economic and Social Aspects of Crime in India* (London: Allen and Unwin, 1934), chap. 9.

28. G. S. Dutt, *A Woman of India: Being the Life of Saroj Nalini*, 2d ed. (London: Hogarth Press, 1929), 77, 88–109, 134.

29. H. D. Sourie, "Homes for the Unattached," *March of India* 4 (1952): 23–24.

30. S. C. Dube, *India's Changing Villages: Human Factors in Community Development* (London: Routledge and Kegan Paul, 1958), 54, 78, plate 12.

31. Madhur Jaffrey, *Climbing the Mango Trees: A Memoir of a Childhood in India* (London: Ebury, 2005), 143.

32. Sharon F. Kemp, "How Women's Work is Perceived: Hunger or Humiliation," in *The Changing Division of Labor in South Asia: Women and Men in India's Society, Economy, and Politics*, ed. James Warner Björkman (Riverdale, MD: Riverdale, 1986), 93.

33. Chatterjee, *Nation and Its Fragments*, 6.

34. Chunilal Bose, *Marriage—Dowry* (Calcutta: Hindu Marriage Reform League, 1914); Patricia Uberoi, "Marriage, Alliance, and Affinal Transactions," in *Family, Kinship and Marriage in India*, ed. Patricia Uberoi (Delhi: Oxford University Press, 1993), 232–33; Ved Mehta, *Face to Face* (Oxford: Oxford University Press, 1978), 96.

35. Bibhutibhushan Bandyopadhyaya, *Aranyak: Of the Forest* (Calcutta: Seagull, 2002), 100.

36. Kees van Dijk, "Pedal Power in Southeast Asia," in *Lost Times and Untold Tales from the Malay World*, ed. Jan van der Putten and Mary Kilcline Cody (Singapore: National University of Singapore Press, 2009), 268–82; David Arnold and Erich DeWald, "Cycles of Empowerment? The Bicycle and Everyday Technology in Colonial India and Vietnam," *Comparative Studies in Society and History* 53 (2011): 971–96.

37. Ronald Ross, *Memoirs* (London: John Murray, 1923), 214. A photograph held by the archives of the London School of Hygiene and Tropical Medicine shows Ross, his wife, and other European cyclists at the Cubon cycle club, c. 1896. The picture appears to show only one Indian, wearing a turban, on the extreme right of the image.

38. Robin M. Le Blanc, *Bicycle Citizens: The Political World of the Japanese Housewife* (Berkeley: University of California Press, 1999).

39. "Reports on the Punjab Disturbances, April 1919," *Parliamentary Papers*, 1920 (Cmd 534), 2.

40. Paul Scott, *The Jewel in the Crown* (London: Granada, 1973), 34, 41.

41. Oscar Jennings, *Cycling and Health*, 2d ed. (London: Iliffe, 1893).

42. H. D. Darukhanawala, *Parsis and Sports and Kindred Subjects* (Bombay: author, 1935), 372–73.

43. S. F. Desai, *Parsis and Eugenics* (Bombay: Mody, 1940), 60–64.

44. Wiebe E. Bijker, *Of Bicycles, Bakelites, and Bulbs: Towards a Theory of Socio-Technical Change* (Cambridge, MA: MIT Press, 1995), 40; Patricia Marks, *Bicycles, Bangs, and Bloomers: The New Woman in the Popular Press* (Lexington: University Press of Kentucky, 1990).

45. Manmohini Zutshi Sahgal, *An Indian Freedom Fighter Recalls Her Life*, trans. Geraldine Forbes (Armonk, NY: Sharpe, 1994), xv, xviii.

46. Mehta, *Face to Face*, 44; Prem Nevile, *Lahore: A Sentimental Journey*, 2d ed. (New Delhi: HarperCollins, 1997), 54.

47. Hemlata C. Dandekar, *Men to Bombay, Women at Home: Urban Influence on Sugao Village, Decca Maharashtra, India, 1942–1982* (Ann Arbor: University of Michigan, 1986), 42.

48. Stree Shakti Sanghatana, *"We Were Making History..." Life Stories of Women in the Telengana People's Struggle* (New Delhi: Kali for Women, 1989), 181, 197.

49. *Guardian* (London), March 12, 2012, 34.

50. *Statesman*, January 15, 1920, 2; ibid., January 20, 1920, 2; ibid., January 22, 1920, 2; ibid., January 25, 1920, 2; ibid., January 27, 1920, 2.

51. India, Home (Establishments), nos. 145–47, January 1915, India Office Records, British Library, London (hereafter cited as IOR); ibid., nos. 8–9, February 1912, IOR; ibid., nos. 211–12, March 1914, IOR.

52. Sumit Sarkar, *Writing Social History* (New Delhi: Oxford University Press, 1997), 176; Dalia Chakrabarti, *Colonial Clerks: A Social History of Deprivation and Domination* (Calcutta: K. P. Bagchi, 2005).

53. Bibhutibhushan Bandyopadhyaya, *Aparajito* (New Delhi: HarperCollins, 1999), 312.

54. *Bombay Chronicle*, July 21, 1920, 2; ibid., December 11, 1926, 2.

55. *Story of the Typewriter, 1873–1923* (Herkimer, NY: Herkimer County Historical Society, 1923), 97.

56. R. V. Marathe, "Lower Middle-Class Life in Bombay," *Social Service Quarterly* 12 (1926): 13, 19.

57. *Statesman*, February 21, 1933, 11.

58. Michael H. Adler, *The Writing Machine* (London: Allen and Unwin, 1973), 40.

59. *CWMG*, 16:93.

60. *Bombay Chronicle*, July 14, 1920, 3.

61. Friedrich A. Kittler, *Gramophone, Film, Typewriter* (Stanford, CA:

Stanford University Press, 1999), 184; R. A. Buchanan, *The Power of the Machine: The Impact of Technology from 1700 to the Present* (London: Penguin, 1994), 175.

62. *Census of India, 1931*, vol. 9, *The Cities of the Bombay Presidency* (Bombay: Central Government Press, 1933), 199; *Census of India, 1931*, vol. 6, *Calcutta* (Calcutta: Central Publications Branch, 1933), 62; *Census of India, 1931*, vol. 14, *Madras*, part 1, *Report* (Madras: Superintendent, Government Press, 1932), 258.

63. Padmini Sengupta, *A Hundred Years of Service: Centenary Volume of the YWCA of Calcutta, 1878–1978* (Calcutta: Gupta, 1987).

64. *Report of the Pauperism Committee* (Calcutta: Bengal Secretariat Press, 1892); *Report of the Calcutta Domiciled Community Enquiry Committee, 1918–19* (Calcutta: Bengal Secretariat Press, 1920).

65. F. DeSouza, *The House of Binny* (Madras: Binny, 1970), 217.

66. Priti Ramamurthy, "All-Consuming Nationalism: The Indian Modern Girl in the 1920s and 1930s," in *The Modern Girl around the World: Consumption, Modernity, and Globalization*, ed. Alys Eve Weinbaum et al. (Durham, NC: Duke University Press, 2008), 153–54.

67. Rohinton Mistry, *Such a Long Journey* (London: Faber and Faber, 1991), 97–99, 173–77, 297; Mulk Raj Anand, *Selected Short Stories*, ed. Saros Cowasjee (New Delhi: Penguin, 2006), 217–18.

Chapter Four

1. Dadabhai Naoroji, *Poverty and Un-British Rule in India* (London: Swan Sonnenschein, 1901), 180; Romesh Dutt, *The Economic History of India*, vol. 1 (London: Routledge and Kegan Paul, 1901).

2. Amales Tripathi, *The Extremist Challenge: India between 1890 and 1910* (Bombay: Orient Longman, 1967), 109–10.

3. R. C. Dutt, December 29, 1906, in *The Indian Industrial Exhibition* (Madras: Srinivasa Varadachari, 1908), 61.

4. Aurobindo Ghosh, *On Nationalism* (Pondicherry: Sri Aurobindo Ashram, 1965), 121.

5. B. P. Wadia, *The Wider Swadeshi Movement* (Adyar: Theosophical Publishing House, 1917).

6. Ananda K. Coomaraswamy, *Art and Swadeshi* (Madras: Ganesh, 1912), 3, 5, 26.

7. Sumit Sarkar, *The Swadeshi Movement in Bengal, 1903–1908* (New Delhi: People's Publishing House, 1973), 134; *Indian Industrial Commission, 1916–18: Report* (Calcutta: Superintendent, Government Printing, India, 1918), 74.

8. Amit Bhattacharyya, *Swadeshi Enterprise in Bengal, 1900–1920* (Calcutta: Mita Bhattacharyya, 1986); Bhattacharyya, *Swadeshi Enterprise in Bengal, 1921–1947: The Second Phase* (Calcutta: Bookland, 1995).

9. *Towards a History of Consumption in South Asia*, ed. Douglas E. Haynes et al. (New Delhi: Oxford University Press, 2010).

10. *Bombay Chronicle*, June 4, 1938, 15.

11. *The Swadeshi Movement: A Symposium* (Madras: G. A. Natesan, 1910), 62–63. An *anna* was a sixteenth part of a rupee.

12. East India (Industrial Commission), *Minutes of Evidence*, vol. 3, *Madras and Bangalore* (London: HMSO, 1919), 404.

13. Sarkar, *Swadeshi Movement in Bengal*, 131.

14. *Census of India 1911*, vol. 6, part 1 (Calcutta, 1913), 73; *Census of India, 1931*, vol. 6, *Calcutta* (Calcutta: Central Publications Branch, 1933), 58; *Census of India, 1931*, vol. 14, *Madras*, part 1, *Report* (Madras: Superintendent, Government Press, 1932), 256; *Punjab District Gazetteer*, vol. 30A, *Lahore District* (Lahore: Superintendent, Government Printing, Punjab, 1916), 154–55.

15. *Report of the Industrial Survey of the Ludhiana District* (Lahore: Superintendent, Government Printing, Punjab, 1942), 73.

16. Pradip Sinha, *Calcutta in Urban History* (Calcutta: Firma K. L. M., 1978), 244.

17. E.g., Somerset Playne, *Southern India: Its History, People, Commerce, and Industrial Resources* (London: Foreign and Colonial Compiling and Publishing, 1915), 162, 181, 664, 689.

18. *Asylum Press Almanack and Directory of Madras and Southern India* (Madras: Chakravarthi, 1928), A-11, 965; *Delhi Directory, 1935* (Delhi, 1935), 62.

19. G. D. Khanolkar, *Walchand Hirachand: Man, His Times and Achievements* (Bombay: Walchand, 1969), 480–81.

20. Mani Bagchi, *Sudhir Kumar Sen: Jiban-charit* (Calcutta: Academy, 1964), 19–20.

21. Prafulla Chandra Ray, *Life and Experiences of a Bengali Chemist* (Calcutta: Chuckervertty, Chatterjee, 1932), 1:404–5.

22. Bagchi, *Sudhir Kumar Sen*, 36–40, 67; *Calcutta Weekly Notes*, 24:155–72.

23. *Bombay Chronicle*, March 12, 1946, 6.

24. *Report of the Indian Tariff Board on the Bicycles Industry* (Bombay: Government Central Press, 1946); *Report of the Indian Tariff Board on the Continuation of Protection to the Bicycle Industry* (Bombay: Manager of Publications, Delhi, 1949).

25. DD/RN/11/1/15, Raleigh Archives, Nottingham Records Office, Nottingham (hereafter cited as RA).

26. S. K. Sen, "The Bicycle Industry and the Second Plan," *Industrial India Annual, 1956*, p. 121, Godrej Archives, Vikhroli, Mumbai (hereafter cited as GA).

27. Bagchi, *Sudhir Kumar Sen*, 89.

28. Mulk Raj Anand, "The Barber's Trade Union," in *Selected Short Stories*, ed. Saros Cowasjee (New Delhi: Penguin, 2006), 7–16.

29. Kathleen Gough, *Rural Society in Southeast India* (Cambridge: Cambridge University Press, 1981), 391.

30. James M. Freeman, *Scarcity and Opportunity in an Indian Village* (Menlo Park, CA: Cummings Publishing, 1977), 99–100.

31. *Bombay Chronicle*, July 16, 1926, 12.

32. *Asylum Press Almanack and Directory of Madras and Southern India* (Madras: Lawrence Asylum Press, 1912), 1256, 1282.

33. India, Home, nos. 3–4, February 1910, India Office Records, British Library, London (hereafter cited as IOR); *The Story of the Typewriter, 1873–1923* (Herkimer, NY: Herkimer County Historical Society, 1923), 129.

34. *Leader*, January 21, 1920, 2; ibid., February 12, 1920, 2.

35. Henry D. Baker, *British India with Notes on Ceylon, Afghanistan, and Tibet* (Department of Commerce, Special Consular Reports, no. 70) (Washington, DC: Government Printing Office, 1915), 316.

36. Friedrich A. Kittler, *Gramophone, Film, Typewriter*, trans. Geoffrey Winthrop-Young and Michael Wutz (Stanford, CA: Stanford University Press, 1999); Rubén Gallo, *Mexican Modernity: The Avant-Garde and the Technological Revolution* (Cambridge, MA: MIT, 2005), chap. 2.

37. S. N. Bhattacharya, *Mahatma Gandhi: The Journalist* (London: Asia Publishing House, 1965), 114.

38. *Malgudi Landscapes: The Best of R. K. Narayan*, ed. S. Krishnan (New Delhi: Penguin, 1992), 8; Ruskin Bond, *The Lamp Is Lit: Leaves from a Journal* (New Delhi: Penguin, 1998), 118–21.

39. Ministry of Commerce, *Report of the Indian Tariff Board on the Machine Tools Industry, 1947* (Simla: Government of India Press, 1918).

40. Arun Joshi, *Lalashri Ram: A Study in Entrepreneurship and Industrial Management* (New Delhi: Orient Longman, 1975), 291.

41. B. K. Karanjia, *Life's Flag Is Never Furled: Godrej, A Hundred Years, 1897–1997* (New Delhi: Penguin, 1997), 1:115.

42. See the correspondence of R. K. Sanjana, Godrej's typewriter sales manager, for 1956–1960, MS 08-01-419-318, GA.

43. Karanjia, *Life's Flag* 1:115–26.

44. Ibid., 114.

45. A. Latifi, *The Industrial Punjab: A Survey of Facts, Conditions and Possibilities* (Bombay: Longmans, Green, 1911), 240.

46. W. H. Moreland, "The Study of Indian Poverty," *Asiatic Review* 16 (1920): 623.

47. *Bombay Department of Agriculture Bulletins*, no. 5 (1886).

48. *Official Report of the Calcutta International Exhibition, 1883–84* (Calcutta: Bengal Secretariat Press, 1885), 1:533–35; Satpal Sangwan, *Science, Technology and Colonialism: An Indian Experience, 1757–1857* (Delhi: Anamika Prakashan, 1991), plates 10–14.

49. *Bombay Department of Agriculture Bulletins*, no. 8 (1887): 10–11.

50. A. C. Chatterjee, *Notes on the Industries of the United Provinces* (Allahabad: Superintendent, Government Press, 1908), 91–95; Shahid Amin, *Sugarcane and Sugar in Gorakhpur: An Inquiry into Peasant Production for Capitalist Enterprise in Colonial India* (Delhi: Oxford University Press, 1984).

51. Latifi, *Industrial Punjab*, 177–79.

52. Bhattacharyya, *Swadeshi . . . Second Phase*, 112.

53. W. R. Aykroyd et al., *The Rice Problem in India* (Calcutta: Thacker, Spink, 1940), 64.

54. Playne, *Southern India*, 613–16.

55. B. P. Adarkar, *Report on Labour Conditions in the Rice Mills* (Delhi: Manager of Publications, 1946), 2, 5.

Chapter Five

1. Timothy Burke, *Lifebuoy Men, Lux Women: Commodification, Consumption, and Cleanliness in Modern Zimbabwe* (Durham, NC: Duke University Press, 1996); Anandi Ramamurthy, *Imperial Persuaders: Images of*

Africa and Asia in British Advertising (Manchester, UK: Manchester University Press, 2003).

2. J. Coatman, *India in 1925–26* (Calcutta: Government of India Central Publication Branch, 1926), 316.

3. Deana Heath, *Purifying Empire: Obscenity and the Politics of Moral Regulation in Britain, India and Australia* (Cambridge: Cambridge University Press, 2010), 158.

4. *Census of India, 1931*, vol. 14, *Madras*, part 1, *Report*, 285.

5. Mulk Raj Anand, "The Cobbler and the Machine," in *Selected Short Stories*, ed. Saros Cowasjee (New Delhi: Penguin, 2006), 64.

6. Bibhutibhushan Bandyopadhyaya, *Aparajito* (New Delhi: Harper-Collins, 1999), 231, 301.

7. Kajri Jain, "New Visual Technologies in the Bazaar: Reterritorialisation of the Sacred in Popular Print Culture," in *Sarai Reader 3: Shaping Technologies* (Delhi: Sarai Programme, 2003), 44–57.

8. *Report of the Press Commission*, part 1 (New Delhi: Manager, Government of India Press, 1954), 80.

9. *CWMG*, 14:5; ibid., 36:272; S. N. Bhattacharya, *Mahatma Gandhi: The Journalist* (London: Asia Publishing House, 1965), 33.

10. For the importance of *swadeshi* advertising, see Amit Bhattacharyya, *Swadeshi Enterprise in Bengal, 1921–1947: The Second Phase* (Calcutta: Bookland, 1995), chap. 8.

11. Harminder Kaur, "Of Soaps and Scents: Corporeal Cleanliness in Urban Colonial India," in *Towards a History of Consumption in South Asia*, ed. Douglas E. Haynes et al. (New Delhi: Oxford University Press, 2010), 246–67; Ranabir Ray Choudhury, *Early Calcutta Advertisements, 1875–1925* (Bombay: Nachiketa, 1992), 469–96.

12. *Madras Mail*, January 2, 1904, 3; *Pioneer*, January 30, 1905, 26.

13. H. D. Darukhanawala, *Parsi Lustre on Indian Soil* (Bombay: printed by author, 1963), 2:xxv.

14. *Bombay Chronicle*, December 11, 1926, 11.

15. Ibid., March 12, 1946, 8.

16. *Illustrated Weekly*, January 8, 1939, 52.

17. Patell to Singer, New York, January 8, 1903, box 89, folder 7, Singer Archive, Wisconsin Historical Society, Madison (hereafter cited as SA).

18. Ray Choudhury, *Early Calcutta Advertisements*, 77; *Bombay Chronicle*, June 2, 1938, 3.

19. Simon Carter, *Rise and Shine: Sunlight, Technology and Health* (Oxford: Oxford University Press, 2007); Vicky Long, *The Rise and Fall of the Healthy Factory: The Politics of Industrial Health in Britain, 1914–60* (Basingstoke, UK: Palgrave, 2010).

20. *Bombay Chronicle*, May 28, 1938, 11.

21. Ibid., June 22, 1938, 11.

22. *Hindu*, January 29, 1950, 6.

23. B. K. Karanjia, *Life's Flag Is Never Furled: Godrej, a Hundred Years, 1897–1997* (Bombay: Godrej, 1997), 1:88, 97.

24. "Sen-Raleigh Industries of India Limited: New 30-Acre Factory at Asansol," DD/RN/11/1/15, Raleigh Archives, Nottingham Records Office, Nottingham (hereafter cited as RA).

25. "Sen-Raleigh Industries of India," [1952], DD/RN/4/3/1/39, RA.

26. *CWMG*, 25:251.

27. Prakash Tandon, *Punjabi Century, 1857–1947* (London: Chatto and Windus, 1963), 162.

28. F. A. Steel and G. Gardner, *The Complete Indian Housekeeper and Cook*, 7th ed. (London: William Heinemann, 1909), 96–97.

29. Ira Klein, "Death in India, 1871–1921," *Journal of Asian Studies* 32 (1973): 639.

30. Patell to Singer, New York, August 22, 1902, box 89, folder 7, SA; September 10, 1902, ibid.; October 15, 1902, ibid.; November 7, 1902, ibid.; September 9, 1903, ibid.; October 29, 1903, ibid.

31. On cycle races and their hazards, see H. D. Darukhanawala, *Parsis and Sports and Kindred Subjects* (Bombay: printed by author, 1935), 376–81; *Bombay Chronicle*, June 2, 1938, 4.

32. Tandon, *Punjabi Century*, 44.

33. *Bombay Chronicle*, December 21, 1933, 3; ibid., June 30, 1938, 5.

34. John Brewer and Roy Porter, introduction to *Consumption and the World of Goods*, ed. John Brewer and Roy Porter (London: Routledge, 1993), 2–3, 7.

35. But for a similar Japanese response, see Harry Harootunian, *Overcome by Modernity: History, Culture, and Community in Interwar Japan* (Princeton, NJ: Princeton University Press, 2000).

36. Amales Tripathi, *The Extremist Challenge: India Between 1890 and 1910* (Bombay: Orient Longman, 1967), 122.

37. Pramatha Nath [Bose], "Western Science from an Eastern View-

point," *Westminster Review* 156 (1901): 219.

38. Pramatha Nath Bose, *National Education and Modern Progress* (Calcutta: Kar, Majumdar, 1921), 19, 24.

39. Cf. Arnold J. Bauer, *Goods, Power, History: Latin America's Material Culture* (Cambridge: Cambridge University Press, 2001), 190–91.

40. L. S. S. O'Malley, *Bengal District Gazetteers: Purnea* (Calcutta: Bengal Secretariat Book Depot, 1911), 110–11.

41. *Census of India 1911*, vol. 1, part 1, 411.

42. Raja Rao, *Kanthapura* (1937) (Westport, CT: Greenwood Press, 1963), 211; M. N. Srinivas, *The Remembered Village* (New Delhi: Oxford University Press, 1972), 234–35.

43. Henry Bruce, "Travancoreans Pounding Rice," *Letters from Malabar and on the Way* (London: George Routledge and Sons, 1909), plate facing 112. Like the machines that replaced them, many preindustrial rice-husking technologies were shared across the rice-producing regions of Asia, as in Vietnam: Maurice Abadie, *Minorities of the Sino-Vietnamese Borderland* (1924), trans. E. J. Tips (Bangkok: White Lotus Press, 2001), 72–73.

44. "JBP," "Rustic Bengal," *Calcutta Review* 59 (1874): 182. For a further description, see A. K. Jameson, "Final Report on the Survey and Settlement Operations in the District of Midnapore," Bengal Revenue Proceedings, nos. 23–24, June 1919, India Office Records, British Library, London (hereafter cited as IOR).

45. Shudha Mazumdar, *A Pattern of Life: The Memoirs of an Indian Woman*, ed. Geraldine H. Forbes (New Delhi: Manohar, 1977), 33.

46. Bibhutibhushan Banerji [Bandyopadhyaya], *Pather Panchali: Song of the Road* (London: Allen and Unwin, 1968), 26, 187.

47. G. N. Gupta, *A Survey of the Industries and Resources of Eastern Bengal and Assam for 1907–08* (Shillong: Eastern Bengal and Assam Secretariat Printing Office, 1908), 75.

48. *Census of India, 1991*, vol. 1, part 1, *Report* (Calcutta: Superintendent, Government Printing, India, 1913), 411, 420; *Census of India, 1931*, vol. 14, *Madras*, part 1, *Report* (Madras: Superintendent, Government Press, 1932), 256.

49. K. Ramiah, *Rice in Madras: A Popular Handbook* (Madras: Superintendent, Government Press, 1937), 89.

50. Kenneth J. Carpenter, *Beriberi, White Rice, and Vitamin B: A Disease, a Cause, and a Cure* (Berkeley: University of California Press).

51. David Arnold, "British India and the 'Beriberi Problem,' 1798–1942," *Medical History* 54 (2010), 293–314.

52. Robert McCarrison and Roland V. Norris, *The Relationship of Rice to Beri-Beri in India* (Calcutta: Thacker, Spink, 1924); R. McCarrison, *Beri-Beri Columbarum* (Calcutta: Thacker, Spink, 1928); W. R. Aykroyd and B. G. Krishnan, "Rice Diets and Beriberi," *Indian Journal of Medical Research* 29 (1941): 551–55.

53. Aykroyd and Krishnan, "Rice Diets," 551.

54. W. R. Aykroyd and B. G. Krishnan, "Diet Surveys in South Indian Villages," *Indian Journal of Medical Research* 24 (1937), 687.

55. Pramatha N. Bose, *Survival of Hindu Civilization*, part II, *Physical Degeneration–Its Causes and Remedies* (Calcutta: Newman, 1921), 48–49.

56. *CWMG* 58:293–94.

57. M. K. Gandhi, *Diets and Diet Reform* (Ahmedabad: Navajivan, 1949), 18, 19, 24–26, 31, 61.

58. *CWMG* 59:404.

59. *CWMG*, 59:226.

60. Prafulla Chandra Ray, *Life and Experiences of a Bengali Chemist* (Calcutta: Chuckerverty, Chatterjee, 1932), 1:387.

61. Hashim Amir Ali, "The Rice Industry in Lower Birbhum: A Survey," *Visva-Bharati Rural Studies* 2 (1934): 35–45. Many studies have noted the negative impact of rice milling, especially on women, e.g., Mukul Mukherjee, "Impact of Modernisation on Women's Occupations: A Case Study of the Rice-Husking Industry of Bengal," *Indian Economic and Social History Review* 20 (1983): 27–45; Smritri Kumar Sarkar, "The Rice Milling Industry in Bengal, 1920–1950: A Case Study of the Impact of Mechanization in the Local Peasant Economy," *Calcutta Historical Journal* 13 (1988–1989): 1–111.

62. Hemlata C. Dandekar, *Men to Bombay, Women at Home: Urban Influence in Sugao Village, Deccan Maharashtra, India, 1942–1982* (Ann Arbor: Michigan University, 1986), 182.

63. Prem Chowdhry, *The Veiled Women: Shifting Gender Equations in Rural Haryana, 1880–1990* (Delhi: Oxford University Press, 1994), 262.

64. Mulk Raj Anand, *Untouchable* (New Delhi: Penguin, 2001), 146.

65. Anand, "Lullaby," in Cowasjee, *Selected Short Stories*, 86–90.

66. Anand, "The Cobbler and the Machine," in ibid., 61–71.

67. Mulk Raj Anand, *Coolie* (New Delhi: Penguin, 1993), 187.

68. Dipesh Chakrabarty, *Rethinking Working-Class History: Bengal,*

1890–1940 (Princeton, NJ: Princeton University Press, 1989); Samita Sen, *Women and Labour in Late Colonial India: The Bengal Jute Industry* (Cambridge: Cambridge University Press, 1999). More generally, see Ahmad Mukhtar, *Factory Labour in India* (Madras: Annamalai University, 1930).

69. B. P. Adarkar, *Report on Labour Conditions in the Rice Mills* (Delhi: Manager of Publications, 1946).

70. *Report on the Working of the Factories Act, Madras, 1915*, p. 14, Madras, Judicial, Government Order 1605 (hereafter cited as GO), June 23, 1916, IOR; *Report on the Working of the Factories Act, Madras, 1941* (Madras: Superintendent, Government Press, 1942), 51. The issue of "loose clothing" was repeatedly identified as a source of accidents in Punjab's cotton-ginning factories: *Annual Report on the Working of the Indian Factories Act in the Punjab, 1925* (Lahore: Superintendent, Government Printing Punjab, 1926), 6.

71. *Annual Report on the Administration of the Factories Act in Bengal, 1936* (Calcutta: Bengal Secretariat Book Depot, 1937), 20–21.

72. E.g., *Report of the Working of the Factories Act, Madras, 1915*, p. 15, Madras, Judicial, GO 1605, June 23, 1916, IOR.

73. Figures calculated from the annual reports of the factory inspectors for Madras and Burma. "Serious injuries" was taken to mean permanent disablement.

74. *Report of the Indian Factory Labour Commission, 1908*, vol. 2, *Evidence* (Simla: Government Central Branch Press, 1908), 297; *Annual Report on the Working of the Indian Factories Act in the Punjab, 1923* (Lahore: Superintendent, Government Printing Punjab, 1924), 3–4.

75. Madras, Revenue (Special), GO 1206, July 2, 1920, Tamil Nadu Archives, Chennai (hereafter cited as TNA).

76. *Annual Report on the Administration of the Factories Act in Bengal, 1935* (Calcutta: Bengal Secretariat Book Depot, 1936), 21.

Chapter Six

1. Satpal Sangwan, *Science, Technology and Colonisation: An Indian Experience, 1757–1857* (Delhi: Anamika Prakashan, 1991), chap. 4.

2. *The Everyday State and Society in Modern India*, ed. C. J. Fuller and Véronique Bénéï (London: Hurst, 2001); Stuart Corbridge, Glyn Wil-

liams, Manoj Srivastava, and René Véron, *Seeing the State: Governance and Governmentality in India* (Cambridge: Cambridge University Press, 2005).

3. Gilbert Slater, *Southern India: Its Political and Economic Problems* (London: Allen and Unwin, 1936), 127.

4. Gregory Houston Bowden, *The Story of the Raleigh Cycle* (London: W. H. Allen, 1975), 131.

5. *The Webbs in Asia: The 1911–12 Travel Diary*, ed. George Feaver (Basingstoke, UK: Macmillan, 1992), 227–29, 244, 312; Edwin S. Montagu, *An Indian Diary* (London: Heinemann, 1930), 235.

6. Benjamin Zachariah, *Nehru* (London: Routledge, 2004), 21.

7. Madras, Public, Government Order 137 (hereafter cited as GO), February 6, 1931, Tamil Nadu Archives, Chennai (hereafter cited as TNA).

8. *CWMG* 41:86.

9. *Hindu*, July 12, 1936, 4.

10. India, Commerce and Industry, nos. 3–4, July 1908, National Archives of India, New Delhi (hereafter cited as NAI).

11. India, Home (Police), nos. 93–94, February 1917, NAI; ibid., nos. 20–21, September 1918, NAI; ibid., nos. 249–50, July 1919, NAI.

12. India, Home (Public), no. 115, 1925, NAI.

13. *Statement Exhibiting the Moral and Material Progress and Condition of India during the Year 1921* (London: HMSO, 1922), 134.

14. *List of Motor Vehicles Registered in Calcutta* (Calcutta: Bengal Secretariat Press, 1922).

15. Katherine Mayo, *Mother India* (London: Jonathan Cape, 1927), 351.

16. India, Home (Public), nos. 175–77, 1984, NAI.

17. In 1928 the Madras police still used the local telephone switchboard without priority over other callers. This was considered unsatisfactory and a separate switchboard was instituted for police use. Madras, Judicial, GO 173, April 1, 1929, TNA. On telephones, see Michael Mann, "The Deep Digital Divide: The Telephone in British India, 1883–1933," *Historical Social Research* 35 (2010): 188–208.

18. Madras, Financial, GO 29, January 12, 1912, TNA; Madras, Judicial, GO 1359, August 24, 1912, TNA.

19. Madras, Judicial, GO 1514, September 27, 1911, TNA.

20. India, Home (Police), no. 130, January 1919, NAI; Madras, Public, GO 443, July 12, 1921, TNA.

21. India, Home (Jails), nos. 26–92, November 1919, NAI; India, Home (Public), 45/30/39, 1939, NAI.

22. Anon., "The Bihar Earthquake: A Personal Narrative," *Asiatic Review* 30 (1934): 276.

23. Shahid Hamid, *Disastrous Twilight: A Personal Record of the Partition of India* (London: Leo Cooper, 1986), 228.

24. India, Home (Municipalities), nos. 31–33, January 1902, NAI.

25. *The Quarterly Indian Army List for January 1, 1911* (Calcutta: Superintendent Government Printing, 1911), 539, 575, 577.

26. *Report of the Indian Police Commission, 1902–03* (Simla: Government Central Printing Office, 1903), 56; Madras, Judicial, GO 47, January 12, 1912, TNA.

27. *Report on the Administration of the Police of the Madras Presidency, 1923* (Madras: Superintendent, Government Press, 1924), 7.

28. Madras, Judicial, GO 173, April 1, 1929, TNA.

29. R. H. Hitchcock, *Peasant Revolt in Malabar: A History of the Malabar Rebellion, 1921* (1925; repr. New Delhi: Usha Publications, 1983), 37–38.

30. [J. Capper], "Indian Jail Industry," *Calcutta Review* 29 (1857): 32–33.

31. Stewart Clark, *History of the Central Prisons of the North-Western Provinces* (Allahabad: Government Press, North-Western Provinces, 1868), 16.

32. India, Home (Judicial), nos. 121–52, October 1882, NAI.

33. Mary Frances Billington, *Woman in India* (New Delhi: Amako, 1973), 246.

34. V. D. Savarkar, *The Story of My Transportation for Life* (Bombay: Sadbhakti Publications, 1950), 112–15, 124.

35. Raghubir Sahai, *Life in Indian Jail* (Allahabad: Allahabad Law Journal Press, 1937), 8.

36. *Report of the United Provinces Jails Reform Committee* (Allahabad: Superintendent, Printing and Stationery, United Provinces, 1938), 18.

37. *Administration Report of the Municipal Commissioner for the City of Bombay, 1906–07* (Bombay: Government Central Press, 1907), vii; *Administration Report of the Corporation of Madras, 1908–09* (Madras: Superintendent, Government Press, 1909), 11; *Annual Report on the Police Administration of the Town of Calcutta, 1907* (Calcutta: Bengal Secretariat Book Depot, 1908), 6; *Annual Report on the Police Administration of the Town of Calcutta, 1913* (Calcutta: Bengal Secretariat Book Depot, 1914), 8.

38. *Report of the Motor Vehicles Insurance Committee, 1936–37* (Delhi: Manager of Publications, 1937), 11–12.

39. *Report on the Administration of the Motor Vehicles Act and the Madras Traffic Rules, 1948* (Madras: Superintendent, Government Press, 1949), 1.

40. David Edgerton, *The Shock of the Old: Technology and Global History since 1900* (London: Profile Books, 2006).

41. Francis Tuker, *While Memory Serves* (London: Cassell, 1950), 15.

42. For the decline of the *jutka*, see *Census of India, 1931*, vol. 14, *Madras*, part 1, *Report* (Madras: Superintendent, Government Press, 1932), 248.

43. *Report on the Municipal Administration of Calcutta, 1918–19* (Calcutta: Corporation Press, 1919), 125; *Report on the Municipal Administration of Calcutta, 1928* (Calcutta: Corporation Press, 1929), 14.

44. *Administration Report of the Corporation of Madras, 1912–13* (Madras: Superintendent, Government Press, 1913), appendix 1, 91; *Administration Report of the Corporation of Madras, 1934–35* (Madras: Superintendent, Government Press, 1935), appendix 1, 119.

45. Ahmad Mukhtar, *Report on Rickshaw Pullers* (Delhi: Managers of Publications, 1946); Rob Gallagher, *The Rickshaws of Bangladesh* (Dhaka: University Press, 1992), 37–39.

46. *Administration Report of the Corporation of Madras, 1931–32* (Madras: Superintendent, Government Press, 1932), 16, 154; *Administration Report of the Corporation of Madras, 1932–33* (Madras: Superintendent, Government Press, 1933), 18.

47. India, Legislative, no. 7, May 1904, appendix U, India Office Records, British Library, London (hereafter cited as IOR).

48. *Report of the Motor Vehicles Insurance Committee, 1936–37* (Delhi: Government of India, Manager of Publications, 1937), 11, 14–15, 53; Ruth Schwartz Cowan, *A Social History of American Technology* (New York: Oxford University Press, 1997), 235.

49. *Report of the Motor Vehicles Insurance Committee, 1936–37*, 17; *Annual Report on the Police of the City of Bombay, 1935* (Bombay: Government Central Press, 1936), 41.

50. *Report on the Administration of the Police of the Madras Presidency, 1937* (Madras: Superintendent, Government Press, 1938), 27.

51. *Statesman*, June 25, 1922, 9; *Bombay Chronicle*, December 15, 1926, 5.

52. *Report on the Administration of the Motor Vehicles Act and the Ma-*

dras Traffic Rules, 1940–41 (Madras: Superintendent, Government Press, 1942), 13.

53. *Report of the Police Administration in the Punjab, 1931* (Lahore: Superintendent, Government Printing, 1932), 10.

54. Indian Chamber of Commerce, *Annual Report of the Committee, 1936* (Calcutta: Indian Chamber of Commerce, 1937), 260–61.

55. *Statesman,* July 21, 1946, 3; Tuker, *While Memory Serves,* 37–39.

56. Henry Knight, *Food Administration in India, 1939–47* (Stanford, CA: Stanford University Press, 1954), chaps. 12 and 14.

57. *Hindu Weekly,* January 29, 1950, 8; ibid., May 14, 1950, 7.

58. *Bombay Chronicle,* December 16, 1947, 5.

59. *Hindu,* March 6, 1946, 4.

60. Notably Ranajit Guha, *Elementary Aspects of Peasant Insurgency in Colonial India* (Delhi: Oxford University Press, 1983).

61. David Arnold, *Colonizing the Body: State Medicine and Epidemic Disease in Nineteenth-Century India* (Berkeley: University of California Press, 1993), 142–44.

62. John Kaye and G. B. Malleson, *History of the Indian Mutiny of 1857–8* (London: William Allen, 1891), 1:134–41; M. N. Das, "Western Innovations and the Rising of 1857," *Bengal Past and Present* 76 (1957): 71–81.

63. Arnold, *Colonizing the Body,* 218–26.

64. *Report on the Administration of the Police of the Madras Presidency, 1920* (Madras: Superintendent, Government Press, 1921), 15–16.

65. Mulk Raj Anand, *Coolie* (Delhi: Penguin, 1993), 226.

66. D. G. Tendulkar, *Gandhi in Champaran* (Delhi: Publications Division, Ministry of Information and Broadcasting, Government of India, 1957), 22, 32, 48.

67. Raja Rao, *Kanthapura* (1937) (Westport, CT: Greenwood Press, 1963), 124.

68. *Hindu,* January 26, 1937, 7; ibid., February 9, 1937, 8.

69. Margaret Bourke-White, *Halfway to Freedom* (New York: Simon and Schuster, 1949), 43, 45.

70. Suranjan Das, "The Crowd in Calcutta Violence, 1907–1926," in *Dissent and Consensus: Protest in Pre-Industrial Societies,* ed. Basudeb Chattopadhyay et al. (Calcutta: Bagchi, 1989), 233–72.

71. Bengal, Judicial (Police), no. 25, August 1898, West Bengal State Archives, Kolkata.

72. *Bombay Chronicle*, April 1, 1946, 5; Tuker, *While Memory Serves*, 38–39, 140.

73. Bourke-White, *Halfway to Freedom*, 16–20.

74. Ibid., 10; Ravinder Kaur, "The Last Journey: Exploring Social Class in the 1947 Partition Migration," *Economic and Political Weekly* 41 (2006): 2221–28.

Epilogue

1. Arundhati Roy, *The God of Small Things* (London: Flamingo, 1997).

2. E. F. Schumacher, *Small Is Beautiful: Economics as If People Mattered* (London: Blond and Briggs, 1973).

3. *Bombay Chronicle*, March 19, 1946, 6.

4. *Selected Works of Jawaharlal Nehru* (New Delhi: Orient Longman, 1976), 8:808–9.

5. Ward Morehouse, "Nehru and Science: The Vision of New India," *Indian Journal of Public Administration* 15 (1969): 497.

6. *Jawaharlal Nehru's Speeches*, vol. 3, *March 1953—August 1957* (Delhi: Publications Division, Ministry of Information and Broadcasting, Government of India, 1958), 4.

7. Dietmar Rothermund, *India: The Rise of an Asian Giant* (New Haven, CT: Yale University Press, 2008), 201; Siddharth Dube, *In the Land of Poverty: Memoirs of an Indian Family, 1947–97* (London: Zed Books, 1998), 13–15.

8. Rothermund, *India*, 161.

Bibliographical Essay

Introduction

The focus of discussion in this book is on India from the 1880s to the 1960s. India is taken as a regionally distinctive example of the character, impact, and signification of technological modernity. But the intention here is also to engage with the more general issue of what constitutes everyday technology, the nature of the interface between technology and society, and the meanings we (or those who precede us in the historical record) assign to modernity.

On the question of what constitutes technology, and for the argument that society and technology are mutually constitutive see Wiebe E. Bijker, Thomas P. Hughes, and Trevor J. Pinch, eds., *The Social Construction of Technological Systems: New Directions in the Sociology and History of Technology* (Cambridge, MA: MIT Press, 1987), an argument further developed in Wiebe E. Bijker and John Law, eds., *Shaping Technology / Building Societies: Studies in Sociotechnical Change* (Cambridge, MA: MIT Press, 1992), and Wiebe E. Bijker, *Of Bicycles, Bakelites, and Bulbs: Towards a Theory of Sociotechnical Change* (Cambridge, MA: MIT Press, 1995). A parallel argument is presented in the editors' introduction to Donald MacKenzie and Judy Wajcman, eds., *The Social Shaping of Technology: How the Refrigerator Got Its Hum* (Milton Keynes, UK: Open University Press, 1985). I have not adopted the distinction sometimes made between a history of "technics" and a history of "technology," but see Donald Cardwell, *The Fontana History of Technology* (London: Fontana, 1994), and the contrasting view of Ruth Schwartz Cowan, "Technology has been a fact of human life as long as there have been human lives" (1), in her book *A Social History of American Technology* (New York: Oxford University Press, 1997).

With few exceptions, such as Joseph Needham's seminal work on

China, until recently most history of technology—at least for the period since 1500 or so—has been a history of the West or the extension of the West into other parts of the globe. Standard works like Cardwell's *Fontana History of Technology*, just cited, make scant mention of technological change in the non-Western world. Work that does relate to the extra-European world often does so in terms of a "technology transfer," recognizing little by way of local adaptation or social reconstitution of imported technologies. Work of this genre that relates to India includes Henry T. Bernstein, *Steamboats on the Ganges: An Exploration in the History of India's Modernization through Science and Technology* (Bombay: Orient Longman, 1960); Daniel R. Headrick, *The Tools of Empire: Technology and European Imperialism in the Nineteenth Century* (New York: Oxford University Press, 1981); Headrick, *The Tentacles of Progress: Technology Transfer in the Age of Imperialism, 1850–1940* (New York: Oxford University Press, 1988); and Roy MacLeod and Deepak Kumar, eds., *Technology and the Raj: Western Technology and Technical Transfers to India, 1700–1947* (New Delhi: Sage, 1995)—though the last of these does contain a more nuanced version of how metropolitan technologies might be locally adapted: see Ian Derbyshire, "The Building of India's Railways: The Application of Western Technology in the Colonial Periphery" (in ibid., 177–215). Of all modern technologies railroads have surely been the most extensively studied in the Indian context: for one of the best examples, see Ian J. Kerr, *Building the Railways of the Raj, 1850–1900* (Delhi: Oxford University Press, 1995), a work that can be usefully supplemented, from a more cultural perspective, by the same author's article "Representation and Representations of the Railways of Colonial and Post-Colonial South Asia," *Modern Asian Studies* 37 (2003): 287–326.

For attempts to rethink the history of technology in modern Asia and Africa, see Francesca Bray's work on China, including "Technics and Civilization in Late Imperial China: An Essay in the Cultural History of Technology," in "Beyond Joseph Needham: Science, Technology, and Medicine in East and Southeast Asia," ed. Morris F. Low, *Osiris* 13 (1998): 11–33, and *Technology and Gender: Fabrics of Power in Late Imperial China* (Berkeley: University of California Press, 1997). For other examples, see Rudolf Mrázek, *Engineers of Happy Land: Technology and Nationalism in a Colony* (Princeton, NJ: Princeton University Press, 2002);

Frank Dikötter, *Things Modern: Material Culture and Everyday Life in China* (London: Hurst, 2007); and Suzanne Moon, *Technology and Ethical Idealism: A History of Development in the Netherlands East Indies* (Leiden, Neth.: CNWS Publications, 2007). An exemplary discussion of small-scale technology in colonial Africa is provided by Nancy Rose Hunt's *A Colonial Lexicon of Birth Ritual, Medicalization, and Mobility in the Congo* (Durham, NC: Duke University Press, 1999).

Issues of what constitutes "modernity" and "the everyday" run through this book and are specifically referenced elsewhere in this essay, but for the problematic nature of the everyday as an analytical concept, see Michel de Certeau, *The Practice of Everyday Life*, trans. Steven Rendall (Berkeley: University of California Press, 1984), and Henri Lefebvre, *Critique of Everyday Life*, vol. 1, *Introduction*, trans. John Moore (London: Verso, 2008), as well as Harry Harootunian, *History's Disquiet: Modernity, Cultural Practice, and the Question of Everyday Life* (New York: Columbia University Press, 2000). More generally on modernity and the everyday, see Arjun Appadurai, ed., *The Social Life of Things: Commodities in Cultural Perspective* (Cambridge: Cambridge University Press, 1986), especially Igor Kopytoff, "The Cultural Biography of Things: Commoditization as Process" (in ibid., 64–91). The contrast with "big technologies" is clear from the essays in "Choosing Big Technologies," ed. John Krige, *History and Technology* 9 (1992).

1. India's Technological Imaginary

There are informative discussions of "imaginaries" in Benedict Anderson, *Imagined Communities: Reflections on the Origin and Spread of Nationalism* (London: Verso, 1983), and Edward Said, *Orientalism* (London: Routledge and Kegan Paul, 1978). Apart from Michael Adas's *Machines as the Measure of Men: Science, Technology, and Ideologies of Western Dominance* (Ithaca, NY: Cornell University Press, 1989), the interpretive possibilities of "imaginaries" have been less well developed in relation to technology and India, but there are elements of an emerging analysis in Markus Daeschel's *The Politics of Self-Expression: The Urdu Middle-Class Milieu in Mid-Twentieth Century India and Pakistan* (London: Routledge, 2005). Gandhi's views on technology are exemplified in his exchanges with Rabindranath Tagore in the early 1920s, for which see Sabyasachi

Bhattacharya, ed., *The Mahatma and the Poet: Letters and Debates between Gandhi and Tagore, 1915–1941* (New Delhi: National Book Trust, 1997). His utopianism and modernity receive interesting treatments in Richard G. Fox, *Gandhian Utopia: Experiments with Culture* (Boston, MA: Beacon Press, 1989), and David Hardiman, *Gandhi in His Time and Ours* (Delhi: Permanent Black, 2003), especially chap. 4.

For "improvement" in colonial ideology, see Richard Drayton, *Nature's Government: Science, Imperial Britain, and the "Improvement" of the World* (New Haven, CT: Yale University Press, 2000), and Joseph Morgan Hodge, *Triumph of the Experts: Agrarian Doctrines of Development and Legacies of British Colonialism* (Athens: Ohio University Press, 2007). For India, see David Arnold, "Agriculture and 'Improvement' in Early Colonial India: A Pre-History of Development," *Journal of Agrarian Change* 5 (2005): 505–25, and for its limitations, see R. J. Henry, "Technology Transfer and Its Constraints: Early Warnings from Agricultural Development in Colonial India," in MacLeod and Kumar, *Technology and the Raj*, 51–77. In part following Bernard S. Cohn's discussion of the "modalities" of colonial knowledge, in his *Colonialism and Its Forms of Knowledge: The British in India* (Princeton, NJ: Princeton University Press, 1996), there have been several contributions to understanding imperial exhibitions and their significance for India, including Carol A. Breckenridge, "The Aesthetics and Politics of Colonial Collecting: India at World Fairs," *Comparative Studies in Society and History* 31 (1989): 195–216; Peter H. Hoffenberg, *An Empire on Display: English, Indian, and Australian Exhibitions from the Crystal Palace to the Great War* (Berkeley: University of California Press, 2001); Gyan Prakash, "Science 'Gone Native' in Colonial India," *Representations* 40 (1992): 153–78; and Lisa Trivedi, *Clothing Gandhi's Nation: Homespun and Modern India* (Bloomington: Indiana University Press, 2007).

On the idea of the "rural idyll," Leo Marx's *The Machine in the Garden: Technology and the Pastoral Ideal in America* (New York: Oxford University Press, 1964) might be read alongside the works of George Birdwood (cited in chapter 1) and E. B. Havell's *Essays on Indian Art, Industry and Education* (Madras: G. A. Natesan, 1910), but also David Ludden, "Craft Production in an Agrarian Economy," in *Making Things in South Asia: The Role of the Artist and Craftsman*, ed. Michael Meister (Philadelphia: University of Pennsylvania, 1988), 103–13. For colonial ethnography, see Nicholas B. Dirks, *Castes of Mind: Colonialism and the Making of Modern India* (Princeton, NJ: Princeton University Press, 2001).

2. Modernizing Goods

The literature of technology transfers (and the social construction of technology) has already been alluded to in the first section of this bibliography. Specific examples include Andrew Godley's essays on sewing machines, notably "The Global Diffusion of the Sewing Machine, 1850–1914," *Research in Economic History* 20 (2001): 1–45, and "Selling the Sewing Machine Around the World: Singer's International Marketing Strategies, 1850–1920," *Enterprise and Society* 7 (2006): 266–313; and Andrew Gordon, *Fabricating Consumers: The Sewing Machine in Modern Japan* (Berkeley: University of California Press, 2012). There are also such standard histories as Robert Bruce Davies, *Peacefully Working to Conquer the World: Singer Sewing Machines in Foreign Markets, 1854–1920* (New York: Arno Press, 1976), and for typewriters, Bruce Bliven's *The Wonderful Writing Machine* (New York: Random House, 1954). Other technologies are less well documented, at least in relation to South Asia, but for milling, see Siok-Hwa Cheng, *The Rice Industry of Burma, 1852–1940* (Kuala Lumpur: University of Malaya Press, 1968), and for sewing machines and gramophones as exemplars of modernity in colonial Ceylon, see Nira Wickramasinghe, *Dressing the Colonised Body: Politics, Clothing and Identity in Sri Lanka* (Hyderabad, India: Orient Longman, 2003), chap. 3.

Of interest in terms of the Indian economy are Christopher John Baker, *An Indian Rural Economy, 1880–1955: The Tamilnad Countryside* (Oxford: Clarendon Press, 1984), which gives a succinct account of the impact of the Depression on Indian agriculture and industry; Clive Dewey, ed., *Arrested Development in India: The Historical Dimensions* (Riverdale, MD: Riverdale, 1988); D. R. Gadgil, *The Industrial Evolution of India in Recent Times, 1860–1939*, 5th ed. (Delhi: Oxford University Press, 1972); and two more recent works, Rajat K. Ray, *Industrialization in India: Growth and Conflict in the Private Corporate Sector, 1914–47* (Delhi: Oxford University Press, 1982), and Tirthankar Roy, *Traditional Industry in the Economy of Colonial India* (Cambridge: Cambridge University Press, 1999).

Aside from newspaper reports and advertisements, almanacs, trade directories, and official statistics on India's maritime trade, a wealth of detail, and interesting photographic evidence, can be found in the volumes compiled by Somerset Playne: *Southern India: Its History, People, Commerce, and Industrial Resources* (London: Foreign and Colonial Compiling and Publishing, 1915); *Bengal and Assam, Behar and Orissa: Their History, People, Commerce, and Industrial Resources* (London: Foreign and Colonial Compiling and Publishing, 1917); and *The Bombay Presidency, the United Provinces, the Punjab etc: Their History, People, Commerce, and Natural Resources* (London: Foreign and Colonial Compiling and Publishing, 1920). Informative, too, are the village studies carried out by Gilbert Slater in Madras between 1915 and 1921, published in his book *Southern India: Its Political and Economic Problems* (London: Allen and Unwin, 1936), and revisited in P. J. Thomas and K. C. Ramakrishnan, eds., *Some South Indian Villages: A Resurvey* (Madras: University Press, 1940). Further information on technological change can be found for other provinces, for example, Malcolm Lyall Darling, *The Punjab Peasant in Prosperity and Debt* (London: Oxford University Press, 1925).

3. Technology, Race, and Gender

More attention has been given to technology in relation to gender than race: for example, Ruth Schwartz Cowan, *More Work for Mother: The Ironies of Household Technology from the Open Hearth to the Microwave* (New York: Basic Books, 1983), but one exception, which combines race, gender, and technology, is Judith A. Carney's *Black Rice: The African Origins of Rice Cultivation in the Americas* (Cambridge, MA: Harvard University Press, 2001). There has been a long tradition, in relation to both colonial and postcolonial India of looking at women in the workplace or in particular occupations, from early studies like Janet Harvey Kelman, *Labour in India: A Study of the Conditions of Indian Women in Modern Industry* (London: Allen and Unwin, 1927), through to more recent works like Maria Mies, *The Lace Makers of Narsapur: Indian Housewives Produce for the World Market* (London: Zed Press, 1982); Hilary Standing, *Dependence and Autonomy: Women's Employment and the Family in Calcutta* (London: Routledge, 1991); and Prem Chowdhry, *The Veiled Women: Shifting Gender Equations in Rural Haryana, 1880–1990* (Delhi: Oxford University Press, 1994). Anthropological works of the postindependence decades by Kathleen Gough, M. N. Srinivas, and William and

Charlotte Wiser offer additional insights into women's engagement with, or exclusion from, modern technology, as do many memoirs and autobiographies of the period. That the gender associations and cultural reception of the sewing machine were different in countries like France can be seen from Karen Offen, "'Powered by a Woman's Foot,' A Documentary Introduction to the Sexual Politics of the Sewing Machine in Nineteenth-Century France, " *Women's Studies International Forum* 11 (1988): 93–101, and Judith G. Coffin, "Credit, Consumption, and Images of Women's Desires: Selling the Sewing Machine in Late Nineteenth-Century France," *French Historical Studies* 18 (1994): 749–83.

On scribes and the scribal tradition in India see the special issue, edited by Rosalind O'Hanlon and David Washbrook, "Munshis, Pandits and Record-Keepers: Scribal Communities and Historical Change in India," *Indian Economic and Social History Review* 47 (2010). The relationship between race and technology has been much less extensively explored, but for interesting work on Anglo-Indians (Eurasians), see Lionel Caplan, "Dimensions of Urban Poverty: Anglo-Indian Poor and their Guardians in Madras," *Urban Anthropology* 25 (1996): 311–49, and his article "Iconographies of Anglo-Indian Women: Gender Constructs and Contrasts in a Changing Society," *Modern Asian Studies* 34 (2000): 863–92. See, too, David Arnold, "European Orphans and Vagrants in India in the Nineteenth Century," *Journal of Imperial and Commonwealth History* 7 (1979): 104–27; Laura Bear, *Lines of the Nation: Indian Railway Workers, Bureaucracy, and the Intimate Historical Self* (New York: Columbia University Press, 2007); and Priti Ramamurthy, "All-Consuming Nationalism: The Modern Girl in the 1920s and 1930s," in *The Modern Girl around the World: Consumption, Modernity, and Globalization*, ed. Alys Eve Weinbaum et al. (Durham, NC: Duke University Press, 2008), 147–73. Otherwise, information about race in relation to technology has to be gleaned from newspapers, business histories, and the official archive.

4. *Swadeshi* Machines

The nature of the *swadeshi* movement in India is discussed in Sumit Sarkar's study *The Swadeshi Movement in Bengal, 1903–1908* (New Delhi: People's Publishing House, 1973); in the earlier work of Amales Tripathi, *The Extremist Challenge: India Between 1890 and 1910* (Bombay: Orient Longman, 1967); and C. A Bayly, "The Origins of Swadeshi (Home Industry): Cloth and Indian Society, 1700–1930," in Arjun Appadurai, *The Social Life of Things*, 285–321. These interpretations have partly been revised by Manu Goswami, *Producing India: From Colonial Economy to National Space* (Chicago: University of Chicago Press, 2004), and recent studies of Indian consumerism, notably the essays in Douglas E. Haynes et al., eds., *Towards a History of Consumption in South Asia* (New Delhi: Oxford University Press, 2010); Abigail McGowan, "All-Consuming Subject? Women and Consumption in Late Nineteenth- and Early Twentieth-Century Western India," *Journal of Women's History* 18 (2006): 31–54; and Judith Walsh, *Domesticity in Colonial India: What Women Learned When Men Gave Them Advice* (New York: Rowman and Littlefield, 2004).

These Indian case studies can be read alongside the wider literature on material culture, consumerism, and advertising in the West, such as Ewen Stuart, *Captains of Consciousness: Advertising and the Social Roots of the Consumer Culture* (New York: McGraw-Hill, 1976); Thomas Richards, *The Commodity Culture of Victorian England: Advertising and Spectacle, 1851–1914* (Stanford, CA: Stanford University Press, 1990); John Brewer and Roy Porter, eds., *Consumption and the World of Goods* (New York: Routledge, 1993); and Timothy Burke, *Lifebuoy Men, Lux Women: Commodification, Consumption, and Cleanliness in Modern Zimbabwe* (Durham, NC: Duke University Press, 1996).

Of equal interest for comparative purposes are studies of technology and modernity which suggest, by contrast, India's rather muted engagement. These include: Rubén Gallo, *Mexican Modernity: The Avant-Garde and the Technological Revolution* (Cambridge, MA: MIT Press, 2005), which has a very different chapter on typewriters (and one on cameras) than can be written for India; Stephen Kern, *The Culture of Time and Space, 1880–1918* (Cambridge, MA: Harvard University Press, 1983), which argues for the dynamic effect of such technological innovations as the telephone, gramophone, automobile, bicycle, cinema, and airplane on the artistic creativity and intellectual life of the period; and two particularly seminal works, Wolfgang Schivelbusch, *The Railway Journey: The Industrialization of Time and Space in the 19th Century* (Berkeley: University of California Press, 1986), and Friedrich A. Kittler,

Gramophone, Film, Typewriter (Stanford, CA: Stanford University Press, 1994). That technological modernity had its critics in the West, too, can be seen from Bernhard Rieger, "'Modern Wonders': Technological Innovation and Public Ambivalence in Britain and Germany, 1890s to 1933," *History Workshop Journal* 55 (2003): 152–76.

5. Technology and Well-Being

On health, masculinity, and effeteness in colonial India, see John Rosselli, "The Self-Image of Effeteness: Physical Education and Nationalism in Nineteenth-Century Bengal," *Past and Present.* 86 (1980): 121–48; Mrinalini Sinha, *Colonial Masculinity: The "Manly Englishman" and the "Effeminate Bengali" in the Late Nineteenth Century* (Manchester, UK: Manchester University Press, 1995); Indira Chowdhury-Sengupta, "The Effeminate and the Masculine: Nationalism and the Concept of Race in Colonial Bengal," in *The Concept of Race in South Asia*, ed. Peter Robb (Delhi: Oxford University Press, 1995), 282–303; and three of my essays, "'An Ancient Race Outworn': Malaria and Race in Colonial India, 1860–1930," in *Race, Science and Medicine, 1700–1960*, ed. Waltraud Ernst and Bernard Harris (London: Routledge, 1999), 123–43; "Diabetes in the Tropics: Race, Place and Class in India, 1880–1965," *Social History of Medicine* 22 (2009): 245–61; and "India and the 'Beriberi Problem,' 1798–1942," *Medical History* 54 (2010): 295–314.

There is an extensive literature on the effects of rice milling in India, especially on women's work, including A. S. Bhalla, "Choosing Techniques: Handpounding v. Machine-Milling of Rice: An Indian Case," *Oxford Economic Papers* 17 (1965): 147–57, and Mukul Mukherjee, "Impact of Modernisation on Women's Occupations: A Case Study of the Rice-Husking Industry of Bengal," *Indian Economic and Social History Review* 20 (1983): 27–45. That the concept of the healthy factory had precedents in Britain (and elsewhere) can be seen, inter alia, from Simon Carter, *Rise and Shine: Sunlight, Technology and Health* (Oxford: Oxford University Press, 2007). The understanding of technology, not in terms of the operation of machines but as technique and ways of doing things, is more evident from Francophone than Anglophone sources, notably Jacques Ellul, *The Technological Society*, trans. John Wilkinson (London: Jonathan Cape, 1965), and Michel Foucault, "Technologies of the Self,'" in *Technologies of the Self: A Seminar with Michel Foucault*, ed. Luther H. Martin et al. (London: Tavistock, 1988), 16–49.

6. Everyday Technology and the Modern State

There has now been an extensive discussion on the nature of the "everyday state" in India; see especially Akhil Gupta, "Blurred Boundaries: The Discourse of Corruption, the Culture of Politics, and the Imagined State," *American Ethnologist* 22 (1995): 375–402; *The Everyday State and Society in Modern India*, ed. C. J. Fuller and Véronique Bénéï (London: Hurst, 2001); *Seeing the State: Governance and Governmentality in India*, ed. Stuart Corbridge et al. (Cambridge: Cambridge University Press, 2005); and Philip Oldenburg, "Face to Face with the Indian State: A Grass Roots View," in *Experiencing the State*, ed. Lloyd I. Rudolph and John Kurt Jacobsen (New Delhi: Oxford University Press, 2006), 184–211. On the question of resistance, see Ranajit Guha, *Elementary Aspects of Peasant Insurgency in Colonial India* (Delhi, Oxford University Press, 1983), and the volumes of *Subaltern Studies* (Delhi: Oxford University Press, 1982), edited by Guha and others, as well as James C. Scott, *Weapons of the Weak: Everyday Forms of Peasant Resistance* (New Haven, CT: Yale University Press, 1985).

Although little has been written about the changing technology of colonial rule, see the essays in MacLeod and Kumar, *Technology and the Raj*, and Michael Mann, "The Deep Digital Divide: The Telephone in British India, 1883–1933," *Historical Social Research* 35 (2010): 188–208. The discussion in this chapter of policing and prisons partly draws on my own work: *Police Power and Colonial Rule: Madras, 1859–1947* (Delhi: Oxford University Press, 1986), and "The Colonial Prison: Power, Knowledge and Penology in Nineteenth-Century India," in *Subaltern Studies 8: Essays in Honour of Ranajit Guha*, ed. David Arnold and David Hardiman (Delhi: Oxford University Press, 1994), 148–87. For the doctrine of laissez-faire and its eventual dissolution, see Sabyasachi Bhattacharya, "Laissez-Faire in India," *Indian Economic and Social History Review* 3 (1965): 1–22; S. Ambirajan, *Classical Political Economy and British Policy in India* (Cambridge: Cambridge University Press, 1978); and Henry Knight, *Food Administration in India, 1939–47* (Stanford, CA: Stanford University Press, 1954). On the many roles of modern technology during partition, see Margaret Bourke-White's *Halfway to Freedom* (New York: Simon and Schuster, 1949); Ravinder Kaul, "The Last Journey: Exploring Social Class in the 1947 Partition Migration," *Economic and Political Weekly* 41 (2006): 2221–28; Penderel Moon, *Divide and Quit: An Eyewitness Account of the Partition of India*, 2nd ed. (Delhi: Oxford University Press, 1998); and Yasmin Khan, *The Great Partition: The Making of India and Pakistan* (New Haven, CT: Yale University Press, 2007).

Index